Aliens and the Antichrist

Aliens and the Antichrist

◆

Unveiling the End Times Deception

When they come, will you be prepared?

John W. Milor

iUniverse, Inc.
New York Lincoln Shanghai

Aliens and the Antichrist
Unveiling the End Times Deception

Copyright © 2006 by John W. Milor

All rights reserved. No part of this book may be used or reproduced by any means, graphic, electronic, or mechanical, including photocopying, recording, taping or by any information storage retrieval system without the written permission of the publisher except in the case of brief quotations embodied in critical articles and reviews.

iUniverse books may be ordered through booksellers or by contacting:

iUniverse
2021 Pine Lake Road, Suite 100
Lincoln, NE 68512
www.iuniverse.com
1-800-Authors (1-800-288-4677)

ISBN-13: 978-0-595-37238-6 (pbk)
ISBN-13: 978-0-595-81636-1 (ebk)
ISBN-10: 0-595-37238-4 (pbk)
ISBN-10: 0-595-81636-3 (ebk)

Printed in the United States of America

Contents

Foreword... ix
OVERVIEW: The Star Trek Version of Reality xiii

Part I THE HEAVENS AS THEY CURRENTLY ARE... 1

CHAPTER 1 Realms of Existence and Their Inhabitants 3
   1. Cursed Domains... 3
   1.1. Cursed Nonangelic Beings............................... 4
   1.2. Cursed Angelic Beings (Fallen Angels).................. 4
   2. Glorified Domains 4
   2.1. Glorified Nonangelic Beings 5
   2.2. Glorified Angelic Beings............................... 5
   3. The Planet Known As Third Heaven 6
   4. Transitional Domains..................................... 7

CHAPTER 2 Overlapping/Coexisting Domains 8
   1. The Prisons in the Lower Regions of the Earth 8
   1.1. Hell—the Places of Departed Human Spirits.............. 9
   1.2. Tartarus and the Abyss—the Places of Nonhuman Spirits . 10
   1.3. The Lake of Fire....................................... 11
   2. The First Heaven 11
   3. The Onion Theory.. 19
   4. Cursed Ground and Holy Ground 25
   4.1. Cursed Ground 26
   4.2. What Does the Bible Say About Ghosts? 27

4.2.1. Glorified Disincarnate Entities. 28
4.2.2. Cursed Disincarnate Entities 30
4.2.3. Interaction with Cursed Disincarnate Entities 33
4.2.4. Advice for People Who See and Hear Spirits of the Deceased 35
4.2.5. Ghostly Visitations of Loved Ones. 37
4.2.6. Beacons of Injustice 38
4.3. Holy Ground 40
5. Interplanetary/Interdimensional Travel 42
5.1. Interplanetary/Interdimensional Travel Conducted by Cursed Beings 43
5.2. Interplanetary/Interdimensional Travel Conducted by Glorified Beings 48

Part II THE HEAVENS OF THE PAST. 55

CHAPTER 3 The Earth Prior to the Creation of Adam and Eve. 61

1. The Flood of Lucifer Mentioned in the Old Testament. 78
2. The Flood of Lucifer Mentioned in the New Testament 93
3. Who Were the Cro-Magnon and Neanderthal? 101
3.1. Three Types of Flesh. 103
3.2. The Inhabitants of Lucifer's Kingdom of the Past. 109
3.3. A New Theory About the Cro-Magnon and Neanderthal. 112
3.4. Does the Bible Give Credence to Atlantis? 115
3.5. The Duration of Lucifer's Kingdom 117
3.6. Deceptions Concerning the Cro-Magnon and Neanderthal 118
4. Lucifer's Dominion beyond the Earth Before and After Adam and Eve 122

Part III THE NEAR FUTURE 127

CHAPTER 4 Could the Rapture of the Church Be a Mass Alien Abduction?. 129

1. The Pretribulation Rapture of the Church Is Consistent with the Heart of God 137
2. The Pretribulation Rapture of the Church Is Consistent with Scripture. 141
2.1. Mixing Up the Second Advent of Christ with the Rapture of the Church. 142

 2.1.1. Details Concerning the Rapture of the Church.................145
 2.1.2. Details Concerning the Second Advent of Christ 148
 2.2. More Than One Resurrection of the Dead, and More Than One Rapture 156
 2.3. Who Hinders the Antichrist?..163
 2.4. What Signs Point to the Nearness of the Rapture of the Church?........165

CHAPTER 5   The Rise of the Antichrist 169
 1. Big Changes Around the Mediterranean Sea....................170
 1.1. The Formation of the Revised Roman Empire....................173
 1.2. The First Conquest of the Antichrist..........................174
 1.3. The Second Conquest of the Antichrist........................179
 1.4. The Destruction of Mystery Babylon............................181
 1.5. Mystery Babylon Identified as the Roman Catholic Church.........187
 1.6. Where Will the United States Stand in Relation to the Antichrist?....189
 1.7. The Antichrist's Frustration with Israel 193
 1.8. The Technology of the Antichrist Kingdom.....................197
 1.9. Could the Antichrist be a Modern-Day Nephilim? 198
 2. The Final Conquest of the Antichrist, and Second Advent of Christ........201

Part IV   THE HEAVENS OF THE FUTURE........ 207

CHAPTER 6   The Millennial Reign of Christ 215
 1. Earth Will Be the Same......................................215
 2. Earth Will Be Different......................................220
 3. The First Assignment Given to the Saints 222
 4. Sin and Death Still Present During the Millennial Reign of Christ........223
 5. The Eternal Future of Life on Earth 224

Part V   THE GRAND CONCLUSION............ 229

APPENDIX   The Gospel...239
Scriptures Referenced in Appendix 243
Notes 251

Foreword

Aliens—who are they, what do they want, and why do they come? Have they been here before? Are they here right now? What role will they play in the future of humanity? What will people think on the day of first contact?

Some people may find this hard to believe, but the Bible answers all of these questions!

Many UFO researchers who quote the Bible refer only to a few passages (Genesis 6, Ezekiel 1:10; 2 Kings 2:11–12, and a few others). They state that UFOs are described in the text, and that is about as far as they go. The majority of them aren't Christians, judging from their definition of God, and they quote the Bible only to support their belief in the existence of extraterrestrial life. While their research provides an excellent resource to prove the existence of extraterrestrial life, their grand conclusions are often vastly flawed. Some of them, such as Erich Von Danekin, in his best-selling book, *Chariots of the Gods*, even conclude that God himself is nothing more than a powerful alien being. Zechariah Sitchen, another best-selling author of numerous books, such as *The 12th Planet*, has a similar conclusion. He sums up his research, based primarily on ancient Sumerian writing, with the conclusion that advanced alien beings are the progenitors of the human race. Even mainstream scientists have jumped on this bandwagon to some extent, suggesting that the origin of life on Earth stems from extraterrestrials, rather than God. Francis Crick and James Watson, the Nobel Prize winners for Physiology and Medicine, are famous for discovering the master molecule that contains the genetic code (DNA). Noting the complexity of DNA has led them to believe it was too complex to be a fluke of nature (evolved from raw non-living material), yet they still couldn't give credit to God for its creation. In the late 1980s, Crick has been noted as "boldly suggesting that the seeds of life on Earth may have been sent here in a rocket launched from some faraway planet by *creatures like ourselves*."[1] This is no new theory. Pre-Socratic Greek philosopher Anaxagoras first espoused the idea of *Panspermia* around 500 BC. After being dormant for over two thousand years, this theory was again revived by scientists in the modern era.[2] With this conclusion, scientists couldn't be more wrong, and part of the goal of this book is to counter this deception. While I believe the Bible documents the existence of alien life, God himself is definitely not an alien—he is the Creator, not a created being.

I believe aliens do exist, however, and the rest of this book is dedicated to discussing their history with Earth, as well as their interactions with humanity in the future ahead of us. The reality we've all thought we knew—the reality we have grown comfortable with—will be completely engulfed into a greater reality that extends into the heavens. The future will be quite interesting, to say the least.

After writing my second book, *Aliens in the Bible*, I focused my attention on writing the book that followed, *The Eaglestar Prophecy*. I consider *The Eaglestar Prophecy* to be one of my most meaningful works; it includes my visions of heaven and God, as well as some of the greatest lessons I've learned in life. When I finished it, I felt my work as an author was complete, but I have continued to correspond via e-mail with many different people about the contents on my Web sites, http://www.AliensAndTheAntichrist.com, http://www.AliensInTheBible.com, and http://www.EaglestarProphecy.com. My books are posted there as free downloads. It seems that people are more interested in paranormal subjects than anything else I write about, but I still enjoy the correspondence, especially when it develops into friendships and the exchange of prayers and encouragement.

By the way, if you haven't figured it out yet, I'm a Christian.

Concerning the correspondence I've had over the past few years, many people have slowly steered my interest back onto the topic of aliens. The discussion centered on matters that I thought I already knew almost everything about. It turned out that I still had a great deal to learn. While I theorized some of the general strategies of Satan concerning the deceptions about to befall Earth, the lacking information I have now gained creates a much clearer picture.

I must admit, I have been a trifle hesitant to undertake writing about aliens again. I've always been a little uncomfortable talking about paranormal subjects because they have an alienating effect (pardon the pun). I sometimes feel uncomfortable talking about paranormal subjects with Christians because the topics are typically categorized as the occult, which most Christians veer away from as a spiritual self-defense mechanism. I don't talk much with non-Christians about these subjects either, because I don't want them to label me as a weirdo. Therefore I've been reluctant to continue writing about aliens, because writing about them sometimes leads to talking about them, but my fixation on the unseen realms (of which aliens/angels are a part) has grown stronger in me with each passing year. Since my chief means of reaching into that realm is through writing, I choose to write, despite the possible ridicule that may befall me.

I once thought of my desire to write about such mysterious things as aliens to be an escape mechanism from various problems I was going through in the past, but I no longer have anything in my life I wish to escape. I love my life! My family is incredible; I love my church; I have a great job; in fact, my life is more blessed and enjoyable than ever before. Nevertheless I still have this peculiar insistence in my spirit to write about these strange things.

Writing about paranormal subjects isn't that bad though. Paranormal subject matter poses some challenges that I'd rather not contend with, but there are also some unique opportunities when a Christian discusses paranormal subjects with non-Christians. The spiritual nature of paranormal subjects makes an easy platform for speaking about spiritual matters of eternal significance. For example, talking about ghosts creates a natural transition to the topic of life after death. In addition, when discussing the existence of aliens, the conversation participants can readily cross over into the realm of the heavens because aliens and angels are the same entities (though not all aliens are angels, all angels can be considered aliens). Without even realizing it, non-Christians sometimes assume a belief system that extends beyond this reality, which requires them to express a certain degree of faith in something they can't see or prove. People willing to take this step of faith are a step closer to believing in God—and ultimately his means of reconciliation through his Son, Jesus Christ.

Many non-Christians also find it intriguing that a Christian would be willing to discuss such topics. Their newly altered perception of Christians helps to bridge the gap between believers and nonbelievers. Aside from these benefits, I also feel called by God to witness to people that flutter about in the New Age movement, primarily dabbling in things they shouldn't be dabbling in. The transition from paranormal subjects to Christianity comes naturally to me; I suspect it's something God has enabled me to do. I sense in the New Age community a seeking heart, which I find particularly easy to reach. I feel that most New Agers are looking for something they know not what, and all it takes is the right person with the right words to reveal to them what they are actually looking for—or rather, *who* they are looking for.

Another positive note concerning writing about paranormal subject matter has to do with the nature of the information I've uncovered. The people of Earth might really benefit from knowing a little more about the amazing history of this planet as it relates to extraterrestrial life, as well as the days that lie ahead. As I write this book, my readership is somewhat small, but I have a feeling that will someday change; and while my audience is probably mostly a secular audience, this will most likely change as well. Therefore this material extends a benefit to both non-Christians and Christians alike, though the non-Christians who will one day become Christians will be the primary benefactors.

My intention in writing this book is not to duplicate everything I covered in *Aliens in the Bible*, although there will be a little bit of overlapping. The only information I wish to impart here is information I haven't provided in *Aliens in the Bible*—and in some cases, topics on which I have changed my views. For this reason, those who have already read *Aliens in the Bible* will be delighted to know that the majority of this information is completely new.

OVERVIEW: The Star Trek Version of Reality

Whoever has read my book *Aliens in the Bible* knows that my interpretation of the definition of heaven comes primarily from my understanding of the Hebrew word "shamayim," found over 650 times in scripture. In most places in scripture, this word is translated as "heavens";[1] it refers to the second heaven (which I will later elaborate on) and is established in the known universe of the sun, moon, and stars, as clearly outlined in Genesis 1:14–18. Anyone with *Strong's Enhanced Lexicon* can look this up.[2]

> Genesis 1:14–18* (bold emphasis and bracketed comments added)
>
> And God said, Let there be **lights [sun, moon, and stars] in the expanse of the heavens** to separate the day from the night. And let them be for signs and for seasons, and for days and years, and let them be **lights in the expanse of the heavens** to give light upon the earth. And it was so. And God made the two great lights—the greater light to rule the day and the lesser light to rule the night—and the stars. And **God set them in the expanse of the heavens** to give light on the earth, to rule over the day and over the night, and to separate the light from the darkness. And God saw that it was good.
>
> *The "lights" are in the "expanse of the heavens," and since these lights are the stars in outer space, then outer space must be the heavens. Heaven is actually composed of three layered spheres of existence (again, I will later elaborate).*
>
> *All quotes in this book are from the King James Version of the Bible, unless otherwise stated.*

Because heaven refers to outer space in most instances, the biblical term "host of heaven," found nineteen times in scripture (eighteen in the Old Testament, one in the New Testament), bears an unfamiliar meaning to most Christians. The term "host of heaven," in short, refers to intelligent life-forms in outer space. Because of this, it is my opinion that *Star Trek* is an excellent depiction of the heavens as outlined in the Bible. The only problem with the *Star Trek* version of reality is the underlying assumption in all *Star Trek* episodes that all the other species in the entire cosmos are just as lost, confused, or indifferent about God as the humans of Earth are. This is only partially true. Most Bible scholars interpret Revelation 12:4 to mean that Satan deceived one-third of the inhabitants of the heavens, while the remaining two-thirds remained faithful to God. Those who are truly faithful to God obviously aren't lost, confused, or indifferent about him.

> Revelation 12:4 (bold emphasis and bracketed comments added)
>
> And his tail drew **the third part of the stars [their inhabitants] of heaven**, and did cast them to the earth: and the dragon stood before the woman who was ready to be delivered, for to devour her child as soon as it was born.

Of the one-third of the host of heaven who are deceived, probably not all of them are deceived to the point that they don't believe God exists at all. Consider the statistics on Earth alone—there is a complex mixture concerning what people believe. Some beings might believe or even know that God exists, yet they don't care; others might be agnostic and indifferent.

Most of the life-forms that exist in the heavens, however, know and dwell in harmony with God. Many of them have even sojourned to his very throne (Nehemiah 9:6; Job 1:6, 2:1, 38:7; Isaiah 6:8, and many others); this is the place referred to in the Old Testament as the highest heaven or heaven of heavens (Deuteronomy 10:14; 1 Kings 8:27; 2 Chronicles 2:6, 6:18; Nehemiah 9:6). In the New Testament, God's throne is termed the "third heaven" (2 Corinthians 12:2–4, which refers to John's vision in Revelation 4:1).

> Nehemiah 9:6
>
> Thou, even thou, art LORD alone; thou hast made heaven, the heaven of heavens, with all their host, the earth, and all things that are therein, the seas, and all that is therein, and thou preserve them all; and the host of heaven worship thee.
>
> Deuteronomy 10:14
>
> Behold the heaven and the heaven of heavens is the LORD's thy God, the earth also, with all that therein is.
>
> 2 Corinthians 12:2–4
>
> I knew a man in Christ above fourteen years ago, (whether in the body, I cannot tell; or whether out of the body, I cannot tell: God knows) such a one caught up to the third heaven. And I knew such a man, (whether in the body, or out of the body, I cannot tell: God knows). How that he was caught up

into paradise, and heard unspeakable words, which it is not lawful for a man to utter.

The highest heaven is clearly described as a glorified planet in the northern hemisphere of the universe. Lucifer led his rebellion into the sides of the north (Isaiah 14:12–14), Mount Zion is in the sides of the north (Psalms 48:2), and promotion comes not from the south, east, or west, but from the Lord, and therefore the north (Psalms 75:6–7).

Isaiah 14:12–14 (bold emphasis added)

How you are fallen from heaven, O Day Star, son of Dawn! How you are cut down to the ground, you who laid the nations low! You said in your heart, I will ascend to heaven; above the stars of God I will set my throne on high; I will sit on the mount of assembly in the **far reaches of the north**; I will ascend above the heights of the clouds; I will make myself like the Most High.

Psalms 48:2* (bold emphasis added)

Beautiful for situation, the joy of the whole earth, is Mount Zion, **on the sides of the north**, the city of the great King.

*Mount Zion is synonymous with Jerusalem, but in this passage of scripture, it refers to the heavenly Jerusalem in the third heaven rather than Jerusalem on Earth, because it's in the sides of the north (unlike the earthly Jerusalem).

Psalms 75:6–7*

For promotion, cometh neither from the east, nor from the west, nor from the south, but God is the judge: He puts down one, and sets up another.

*Promotion doesn't come from the east, west, or south, yet the north is conspicuously left out.

The majority of the host of heaven understand their precise purpose for existing, as defined for them by God himself. However, as far as I am aware, no such species has ever been depicted on a single episode of *Star Trek*. Aside from that, good ol' Gene (Gene Roddenberry, the author of *Star Trek*), nearly got it right, in my opinion.

Expounding on what I'm calling the *Star Trek* version of reality, I will outline what I think the cosmos is truly like, interlacing my synopsis with scriptural support. I will begin with a brief description of how the heavens currently are. I will then divulge the past, even extending back in time to the era known as the dispensation of angels (the time before Adam and Eve were created). Finally I will conclude my description of the greater reality by looking forward in time to events that are about to come upon the Earth, and I will explain how these events extend beyond the Earth and back into the heavens.

Part I
The Heavens as They Currently Are

The expanse of the universe consists of an incomprehensible number of stars. Estimates vary extensively, ranging from 400 billion stars in the Milky Way (multiplied by an additional estimated 80 million other galaxies) to 70 sextillion (seven followed by twenty-two zero's) or more that we know of.[1] Suffice it to say, even the smallest estimate accounts for a vast number of solar systems.

(Note that the above source references are specific Web pages that are subject to change. If these Web pages change, a good way to find them is to drop back to the root directory of the Web site and look through the links on the main page. For example, if the Web page "http://ask.yahoo.com/20010810.html" no longer exists, drop back to http://ask.yahoo.com, and begin searching through links from there. Also, another trick that works is to perform a search for the title of a Web page in a search engine like http://www.yahoo.com, or http://www.google.com.)

Continuing on, these solar systems mentioned above contain multidimensional orbiting planets, which are planets in the universe (or multiverse) that coexist within multiple dimensions, just as Earth exists in multiple dimensions (explained later). Many of these solar systems contain life-forms that, like the planets they inhabit, span multiple dimensions. (Note that I consider the scientific term of "dimensions" to be equivalent to the biblical "levels of glory.") If this terminology is somewhat confusing at this point, there's no need for concern because the pages that follow will explain everything as clearly as possible.

1

Realms of Existence and Their Inhabitants

1. Cursed Domains

As mentioned already, one-third of the worlds that contain life in the cosmos are deceived, and therefore similar to Earth in many respects; they are cursed worlds inhabited by cursed beings. Things such as crime, pollution, and diseases are common.

Two types of cursed beings dwell among the cursed domains of the universe. First are those who have been translated to angelic status. These are immortal beings with bodies that have been translated into a glorified state of being that no longer reproduce (Matthew 22:30; Mark 12:25; Luke 20:35). Second are those who have not been translated into angelic status yet. These are beings who are like humans in the respect that they are physically mortal, and they might be as lost and confused about God as the humans of Earth are.

Matthew 22:30 (bracketed comments added)

For in the resurrection they neither marry, nor are given in marriage [and therefore don't reproduce], but are as the angels of God in heaven.

Mark 12:25

For when they shall rise from the dead, they neither marry, nor are given in marriage; but are as the angels, which are in heaven.

Luke 20:35–36

But they which shall be accounted worthy to obtain that world, and the resurrection from the dead, neither marry, nor are given in marriage: Neither can they die any more: for they are equal unto the angels; and are the children of God, being the children of the resurrection.

1.1. Cursed Nonangelic Beings

Those host of heaven (including humans, because they are called hosts in Exodus 12:41, and many other scriptures) who have never reached angelic status are doomed to die physically, but there may be hope for these beings to repent and be saved. The humans of Earth fall into this category, but they're not alone. Jesus may have spread his word to other nonangelic beings in the universe, one way or another. It only makes sense that he would if they exist, and I think the Bible infers that they do. Again, I will talk more about this later.

> Exodus 12:41
>
> And it came to pass at the end of the four hundred and thirty years, even the selfsame day it came to pass, that all the hosts of the LORD went out from the land of Egypt.

1.2. Cursed Angelic Beings (Fallen Angels)

As for those translated into angelic status, their situation is worse than those who were never translated. These extremely powerful beings knew God intimately and spoke with him face to face, yet they turned away from him in disobedience. Because of their high level of accountability, and the intimacy with which they knew God, they have sinned to the degree that God will never forgive them. Their curse will eventually result in the eternal destruction of their immortal bodies and spirits (Revelation 20:15).

> Revelation 20:15
>
> And whosoever was not found written in the book of life was cast into the lake of fire.

2. Glorified Domains

The other two-thirds of the worlds containing life in the cosmos (Revelation 12:4) are also similar to Earth in some respects. They are physical planets with all the features of physical planets (explained more in depth in successive chapters), like various terrains, such as mountains, valleys, trees, a sky, clouds, and so forth, which are all present. Because they are glorified worlds, however, their differences from Earth have to do with how much better they are than Earth. Glorified worlds are much more beautiful than Earth, and they don't contain any death, disease, crime, or pollution. They furthermore serve as the homes of immortal beings who have supernatural power as a common trait (1 Corinthians 13:8–10).

> 1 Corinthians 13:8–10*
>
> Charity never fails: but whether there are prophecies, they shall fail; whether there are tongues, they shall cease; whether there is knowledge, it shall vanish away. For we know in part, and we prophesy in part. But when that which is perfect is come, then that which is in part shall be done away.
>
> *This passage speaks of the glorified immortal nature that saved humans will one day have, because we will be perfect. Currently, all human gifts are only partial gifts (in part). In the future, however, people will be like the angels who are in heaven—angels who have not sinned are perfect and imbued with supernatural power as a common trait.*

2.1. Glorified Nonangelic Beings

Just as there are two classes of cursed beings, there are two classes of glorified beings. The first class of glorified beings inhabiting some worlds comprises those beings who are in transitional states, as Adam and Eve were. In essence they have not yet been translated to angelic status, but they are not cursed with death either. One primary characteristic that distinguishes them from angels is that they are required to procreate in order to populate their planets (Genesis 1:28). Humans prior to the fall were in this transitional state. At some point in their existence, if beings in this glorified category maintain their obedience to God, they will eventually be translated to angelic status. As for humanity, this was the original plan, and it will still happen, for those who have repented (even though they currently aren't glorified). This opportunity exists only because of the work of Jesus Christ (Matthew 22:30; Mark 12:25; Luke 20:35; Revelation 21:16–17).

2.2. Glorified Angelic Beings

Also inhabiting some of the two-thirds of the glorified worlds are beings who have been translated into angelic status. Most, if not all, glorified worlds contain both classes of glorified beings. I say this based on the Bible's description of Earth in the future. When Earth is glorified once again, it will contain both angels and natural generations of people. Glorified saints will be angels (Revelation 21:16–17), and natural generations of people who survive the days of the Antichrist will also be on the Earth forever as well (Isaiah 2:1–2, 11:6–7, 65:25; Zechariah

14:16–21; Revelation 11:15, 20:1–10; Matthew 25:31–46). Actually translation probably does not occur for all the members of a given species at the same time. Massive group translations are only on rare occasions, such as the rapture of the church.

Concerning glorified angelic beings, it is my assumption that they are generally much more powerful and advanced than nonangelic beings. This assumption only comes from common sense and from reading about the abilities of angels as compared to those of Adam and Eve, but it isn't necessarily a scripturally provable fact. Simply consider that an advanced species, the members of which work together in harmony for millions or even billions of years accumulating wisdom and knowledge, is going to achieve more of its cognitive and spiritual potential than a species that has only existed for a few thousand years. There may be variables that can distort this conclusion, such as the sharing of knowledge between species; generally, however, the more glory a being has been given from God, the more powerful it is in every conceivable facet.

3. The Planet Known As Third Heaven

In the northern hemisphere of the universe, there exists a certain planet where God has established his throne (Lucifer led his rebellion into the sides of the north—Isaiah 14:12–14; mount Zion is in the sides of the north—Psalms 48:2; promotion comes not from the east, west, or south, but from the Lord and therefore the north—Psalms 75:6–7). To suggest that this highest heaven where God dwells is not a planet is to suggest that it is an eternally extending flat surface stretching out into an infinite horizon, even though in every other respect, according to descriptions given in scripture, it resembles a planet (with mountains, trees, rivers, cities, and a sky with clouds). Why the human mind doesn't naturally connect the concept of heaven with that which is known, as it is stated in Romans 1:20, is a strange mystery. The Bible describes the third heaven as a physical realm very much like Earth in every sense conceivable. It's absurd to conclude that it is an infinite, flat plane, rather than a glorified planet.

Romans 1:20–21 (bold emphasis and bracketed comments added)

For the invisible things [such as the planet heaven] of him from the creation of the world are clearly seen, being understood by the things that are made [things on the Earth], even his eternal power and Godhead; so that they are without excuse: because that, when they knew God, they glorified *him* not as God, neither were thankful; but became vain in their imaginations, and their foolish heart was darkened.

This planet known as the third heaven contains the highest level of glory in existence. The city where God's throne is currently located is known as New Jerusalem. Every description of this realm fits that of a glorified planet, which is described in detail in scripture. (The third heaven will be covered more in depth throughout this book.)

4. Transitional Domains

Just as living beings go through transitions of glory, planets go through transitions of glory. I say this because of the various stages of glory Earth has gone through, and is yet to go through. Earth once housed a parcel of glorified territory known as the Garden of Eden, but after the fall of Adam and Eve, it became cursed. (Earth's history with the Garden of Eden actually predates Adam and Eve, which is a point of view I adopted after I wrote *Aliens in the Bible*. I will cover this information later in this book.)

Earth has been cursed since the fall of Adam and Eve, but not all hope is lost: Earth will not only contain a place of paradise, as it once did, but also the paradise it will contain in the future will be even better than the Garden of Eden that once was. Earth will be given the honor of housing the capital city of the entire universe, complimented by the very throne of God himself.

2

Overlapping/Coexisting Domains

When I first described the heavens as they currently are, I briefly mentioned the concept of the multidimensional planets in a multidimensional universe. According to scripture, Earth houses more than one level of existence, and these levels of existence vary in their degree of glory.

1. The Prisons in the Lower Regions of the Earth

Scripture mentions the existence of at least four regions within the Earth:

- Hell, the two places of departed spirits: "Hades" in Greek and "Sheol" in Hebrew, both of which are translated as "hell" and consisted of two regions, a place of torment (Matthew 16:18; Luke 16:19–31) and a place called paradise, or Abraham's bosom (Luke 16:19–31; 23:43).
- Tartarus, where fallen angels who sinned in Genesis 6 are held (1 Peter 3:19; 2 Peter 2:4; Jude 6–7).
- The Abyss, referred to as "the deep" (Luke 8:26–31; Romans 10:7), and "bottomless pit" (Revelation 9:1–3, 9:11, 11:7, 17:8; 20:1–10).

Scripture also mentions a place of eternal destruction, called the lake of fire, which the above four mentioned places are thrown into after Judgment Day (Revelation 20:14). It could be that the lake of fire is in the center of Earth, but I'm doubtful of this, since it is never mentioned in reference to the center of Earth. Other parts of scripture suggest that the lake of fire will not be located within Earth; I will disclose more information on this in successive chapters.

Revelation 20:14

And death and hell were cast into the lake of fire. This is the second death.

1.1. Hell—the Places of Departed Human Spirits

Prior to the resurrection of Christ, there was a place in the center of Earth called hell ("Hades" in Greek, "Sheol" in Hebrew). It's still there, but its structure is most likely different now than it was before the resurrection of Christ. In the Old Testament times, Hades/Sheol was a prison for *all* departed human spirits. One part of it was known as Abraham's Bosom; it was also called Paradise. It was a blissful place of peace, reserved for the saints of the Old Testament times, but it was still a prison separating them from God. The other part of Hades/Sheol was a place of burning fire, with varying degrees of torment. A chasm separated the paradise portion of hell from the torment region. This structure of Hades/Sheol was altered after Christ died on the cross. When he entered the paradise portion of hell, he freed all the spirits trapped there (Luke 23:42–43; 2 Corinthians 5:8).

> Luke 23:42
>
> And he said unto Jesus, Lord, remember me when thou comest into thy kingdom. And Jesus said unto him, Verily I say unto thee, today you will be with me in paradise.
>
> 2 Corinthians 5:8*
>
> We are confident, I say, and willing rather to be absent from the body, and to be present with the Lord.
>
> *Being in the presence of the Lord while apart from the body is a blessing that the Old Testament prophets Abraham and Samuel did not enjoy until Christ fulfilled his mission on the cross (Luke 16:25; 1 Samuel 28:11–15).*

Now, in New Testament times, all departed spirits faithful to God get to go to the highest heaven to be with God (2 Corinthians 5:8) and await a reunification with their physical bodies. The structure of Hades/Sheol was therefore altered after the crucifixion of Christ. If the paradise portion of hell still exists, it's either empty, or it was taken over by beings from the torment region of hell or fallen angels who still inhabit the first and/or second heavens. I personally lean in favor of the idea that it was taken over by fallen angels in the first and/or second heaven. If beings from the torment region of hell could have traversed the chasm to get there, they would've done so long before Christ entered the area, and the place known as Abraham's Bosom wouldn't have been such a delightful place to be. In any case, if this place still exists, and evil beings have taken it over, it would explain why non-Christians who have had near-death experiences frequently describe a place of paradise, which motivates them to perpetuate the false doctrine of the nonexistence of a place of burning torment after death. Noted physician, lecturer, and authority on near-death experiences, Raymond Moody, Jr., Ph.D., MD, documents in his research, over 100 cases of near-death experiences, *all* of which were extremely positive. I find it more likely to suggest some kind of deception is at work here, rather than postulate that *all* of the near-death cases he documented just happened to be Christians.

1.2. Tartarus and the Abyss—the Places of Nonhuman Spirits

Scripture mentions a place known as Tartarus, which is described as a place specifically designed to hold the spirits of the fallen angels who sinned before and after the flood of Noah; scripture has no record of humans or demons ever going to this place of torment (note that demons are not the same as fallen angels). Scripture also holds no record of humans going to a place referred to as the Abyss; the Abyss is also called the bottomless pit. The Abyss is like Tartarus in that it was designed for fallen angels (Revelation 9:3–11), but unlike Tartarus, it also contains disincarnate pre-Adamite spirits as well. Both the Nephilim (only half-human—Genesis 6) and other beings of an earlier creation (inhabitants of the pre-Adamite civilization), are held there as well, and these beings will be discussed later in this book.

> Revelation 9:3–11 (bold emphasis added)
>
> And there came out of the smoke locusts upon the Earth: and unto them was given power, as the scorpions of the Earth have power. And it was commanded them that they should not hurt the grass of the Earth, neither any green thing, neither any tree; but only those men which have not the seal of God in their foreheads. And to them it was given that they should not kill them, but that they should be tormented five months: and their torment was as the torment of a scorpion, when he strikes a man. And in those days shall men seek death, and shall not find it; and shall desire to die, and death shall flee from them. And the shapes of the locusts were like unto horses prepared unto battle; and on their heads were as it were crowns like gold, and their faces were as the faces of men. And they had hair as the hair of women, and their teeth were as the teeth of lions. And they had breastplates, as it were breastplates of iron; and the sound of their wings was as the sound of chariots of many horses running to battle. And they had tails like unto scorpions, and there were stings in their tails: and their power was to hurt men five months. **They had a king over them, who is the angel from the bottomless pit.** His

name in the Hebrew tongue is Abaddon, but in the Greek tongue hath his name Apollyon.

1.3. The Lake of Fire

The most terrible place of all is the lake of fire. There are currently no spirits in this particular place because scripture states that they will be cast there in the future. The first beings thrown there will be the false prophet and the Antichrist—who is referred to as the beast (Revelation 19:20). Then, on the Day of Judgment following the thousand-year reign of Christ, Satan and everyone else who never believed or accepted God's invitation to salvation will be thrown there (Revelation 20:10–15).

While Hades/Sheol, Tartarus, and the Abyss are all temporary regions in inner Earth that will one day pass away, the lake of fire will exist forever. The lake of fire is the final location of torment for all unfaithful spirits, both human and non-human (Revelation 20:10–15).

2. The First Heaven

Aside from the various realms existing within Earth, scripture also mentions the existence of a realm in the immediate sky surrounding Earth. This place, which geographically is the upper atmosphere of Earth, is referred to indirectly as the first heaven. Simply look up the word "heaven" in *Strong's Enhanced Lexicon*, and you will find "sky" as one of the definitions.[1] Many scriptures that refer to heaven are clearly speaking of the sky. People were looking into the sky when the Holy Spirit descended upon Jesus as a dove (Mark 1:10 and others); Jesus was physically taken up into the sky in a cloud, which the angels referred to as heaven (Acts 1:9–11); and New Jerusalem will descend from the sky, which is called heaven (Revelation 21:2). Many other scriptures refer to the clouds as heaven (Genesis 7:23, 8:2; Jeremiah 51:16, and others).

Mark 1:10–11

And straightway coming up out of the water, He saw the heavens opened, and the Spirit like a dove descending upon him: And there came a voice from heaven, saying, Thou art my beloved Son, in whom I am well pleased.

Acts 1:9–11

And when He had spoken these things, while they beheld, He was taken up; and a cloud received him out of their sight. And while they looked steadfast toward heaven as He went up, behold, two men stood by them in white apparel; which also said, Ye men of Galilee, why stand ye gazing up into heaven? This same Jesus, who is taken up from you into heaven, shall so come in like manner as ye have seen him go into heaven.

Revelation 21:2

And I John saw the holy city, New Jerusalem, coming down from God out of heaven, prepared as a bride adorned for her husband.

Genesis 7:23 (bold emphasis added)

And every living substance was destroyed which was upon the face of the ground, both man, and cattle, and the creeping things, **and the fowl of the heaven**; and they were destroyed from the earth: and Noah only remained alive, and they that were with him in the ark.

The first heaven is not to be confused with the second heaven. Second heaven is outer space (another definition given in *Strong's Enhanced Lexicon*, clearly derived from the context of celestial spheres—Genesis 1:15–20, 22:17; Deuteronomy 4:19; Isaiah 14:12–14; Psalms 8:3, 148:3–4; Ezekiel 32:8).[2] The first heaven shouldn't be confused with the third heaven either. The third heaven is also called the highest heaven (Deuteronomy 10:14; 1 Kings 8:27; 2 Chronicles 2:6, 6:18; Nehemiah 9:6; Psalms 115:6; 2 Corinthians 12:2; Revelation 8:10 and many others). The third heaven is the glorified planet where God's throne is, located in the northern hemisphere of the universe (Isaiah 14:12–14; Psalms 75:6–7).

Concerning the first heaven, scripture states that Satan is currently in charge of this region of Earth (with God-imposed limitations). Satan is known as the god of this world (2 Corinthians 4:4), the prince of the power of the air (Ephesians 2:1–3), and the prince of this world (John 12:31). I consider this fact alone to be sufficient evidence suggesting that many alien abductions are conducted by fallen angels who are in the first heaven—right here on Earth.

Delving farther into this realm, the book of Daniel mentions a being known as the "prince of Persia," a fallen angel who hindered the archangel Gabriel for twenty-one days when Gabriel set out on his mission to deliver a message to the prophet Daniel (Daniel 10:5–21). One can even picture the event: Gabriel traveled from the third heaven (the planet of God's throne), through the second heaven (outer space), passing through various dimensions of glory as he went, then encountered resistance once reaching the first heaven (the outer atmosphere of Earth). The archangel Michael ("Daniel's prince," and therefore the prince of Israel—Daniel 10:21) assisted him against this resistance, leading an angelic procession from the region in the first heaven located over Israel, where he rules, into the region over Persia, where Daniel was located. Michael battled against the prince of Persia and other satanic beings in authority over the region of Persia in order to penetrate the first heaven to get to Daniel. By the time the archangel Gabriel was finally able to breach the enemy's blockade and enter our current dimension, twenty-one days had passed. (This sounds sort of like *Star Wars*; those who wonder why prayers take a little while to be answered sometimes can find a clue from these passages.)

Daniel 10:5–21

Then I lifted up mine eyes, and looked, and behold a certain man clothed in linen, whose loins were girded with fine gold of Uphaz: His body also was like the beryl, and his face as the appearance of lightning, and his eyes as lamps of fire, and his arms and his feet like in color to polished brass, and the voice of his words like the voice of a multitude. And I Daniel alone saw the vision: for the men that were with me saw not the vision; but a great quaking fell upon them, so that they fled to hide themselves. Therefore I was left alone, and saw this great vision, and there remained no strength in me: for my comeliness was turned in me into corruption, and I retained no strength. Yet heard I the voice of his words: and when I heard the voice of his words, then was I in a deep sleep on my face, and my face toward the ground. And, behold, an hand touched me, which set me upon my knees and upon the palms of my hands. And he said unto me, O Daniel, a man greatly beloved, understand the words that I speak unto thee, and stand upright: for unto thee am I now sent. And when he had spoken this word unto me, I stood trembling. Then said he unto me, Fear not, Daniel: for from the first day that thou didst set your heart to understand and to chasten thyself before thy God, thy words were heard, and I am come for thy words. But the prince of the kingdom of Persia withstood me one and twenty days: but, lo, Michael, one of the chief princes, came to help me; and I remained there with the kings of Persia. Now I am come to make thee understand what shall befall thy people in the latter days: for yet the vision is for many days. And when he had spoken such words unto me, I set my face toward the ground, and I became dumb. And, behold, one like the similitude of the sons of men touched my lips: then I opened my mouth, and spoke, and said unto him that stood before me, O my lord, by the vision my sorrows are turned upon me, and I have retained no strength. How can the servant of this my lord talk with this my lord? For as for me, straightway there remained no strength in me, neither is there breath left in me. Then there came again and touched me one like the appearance of a man, and he strengthened me, and said, O man greatly beloved, fear not: peace be unto thee, be strong, yea, be strong. And when he had spoken unto me, I was strengthened, and said, Let my lord speak; for thou hast strengthened me. Then said he, Do you know wherefore I come unto thee? And now will I return to fight with the prince of Persia: and when I am gone forth, lo, the prince of Grecia shall come. But I will show thee that which is noted in the scripture of truth: and there is none that holds with me in these things, but Michael your prince.

An interesting thing to note in this passage is that the Grecian Empire was not yet in power on Earth at the time that scripture was written. Gabriel specifically stated that the battle Gabriel and Michael were to fight when returning to heaven would be against the prince of Persia, so that the prince of Grecia could take his place (Daniel 10:10). This indicates that the Persian Empire (ruled by the evil "prince of Persia") came into power in a higher dimension on Earth prior to that power making its manifestation in our current dimension. It was also removed from power and replaced with the new authority called the "prince of Grecia." Therefore all earthly empires are a subset of the higher level of existence in the first heaven. Jesus spoke of this relationship between the first heaven and the Earth, when he gave the apostles the keys to the kingdom of heaven (Matthew 16:19). (Incidentally, the prince of Grecia is the same fallen angel, the "Destroyer," who will be released from the Abyss in the end times in order to assist the Antichrist.)[3]

Daniel 10:20

Then said he, Do you know wherefore I come unto thee? And now will I return to fight with the prince of Persia: and when I am gone forth, lo, the prince of Grecia shall come.

Matthew 16:19

And I will give unto thee the keys of the kingdom of heaven: and whatsoever thou shalt bind on earth shall be bound in heaven: and whatsoever thou shalt loose on earth shall be loosed in heaven.

While on the subject of the ancient Grecian Empire rising to power, I find it intriguing to note that Alexander the Great may have received some divine help during his siege of Tyre in 332 BC, explained in the excerpt below.

Quoting Giovanni Gustavo Droysens Storia di Alessandro il Grande, the erudite Italian Alberto Fenoglio, writes in CLYPEUS Anno 111, No 2, a startling revelation, which we now translate.

"The fortress would not yield, its walls were fifty feet high and constructed so solidly that no siege-engine was able to damage it. The Tyrians disposed of the greatest technicians and builders of war-machines of the time and they intercepted in the air the incendiary arrows and projectiles hurled by the catapults on the city. One day suddenly there appeared over the Macedonian camp these flying shields, as they had been called, which flew in triangular formation led by an exceedingly large one. The others were smaller by almost one-half. In all, there were five."[4]

The unknown chronicler narrates that they circled slowly over Tyre while thousands of warriors on both sides stood and watched them in astonishment. Suddenly from the largest shield came a lightning-flash that struck the walls; these crumbled and other flashes followed. The walls and towers dissolved, as if they had been built of mud, leaving the way open for the besiegers, who poured like an avalanche through the breeches. The 'flying shields' hovered over the city until it was completely stormed, then they very swiftly disappeared aloft, soon melting into the blue sky.[5]

Could this account be a glimpse of the battle that the archangel Gabriel mentioned to the prophet Daniel, spilling over into our dimension?

The apostle Paul refers to the hierarchy of the first heaven (and second heaven as well, but not the third heaven in this particular reference—because the third heaven does not contain satanic rulers anymore), with its satanic inhabitants in their positions of power, in Ephesians 6:12. The first heaven has an established governmental hierarchy, much the same way Earth is ruled in our current dimension. Some regions have faithful angels who rule over them (such as Israel, and probably the United States, indicated by our alliance with Israel), while fallen angels have power over other regions, just as in our dimension, where evil dictators rule some countries, and righteous leaders rule other countries.

The Old Testament Septuagint (Greek translation of the Hebrew Bible that Jesus and the apostles quoted from roughly 80 percent of the time) also refers to the hierarchy of the first heaven, when it states that God distributed the regions of the Earth according to the "sons of God" (angelic beings). While this translation is not seen in the King James Version of the Bible, it is clearly stated in the English Standard Version of the Bible, which is consistent with the Dead Sea Scrolls in this passage.[6] The excerpt below is from the English Standard Version.

> Deuteronomy 32:8–9 (bracketed comments added)
>
> When the Most High gave to the nations their inheritance, when he divided mankind, he fixed the borders of the peoples according to the number of the sons of God [angels, both faithful and fallen]. But the LORD's portion is his people, Jacob his allotted heritage.

God decided to do this territorial distribution to various angels in the first heaven (some faithful, some fallen), in response to the rebellious nature of the people of Earth. Again, as I mentioned above, Jesus stated to the apostles that the things that are bound on Earth are bound in heaven, and the things loosed on Earth, are loosed in heaven. Therefore, the actions of the people of the Earth determine the kind of authority that will be established over them in the first heaven. If they turn from God, he turns them over to corrupt authorities in the first heaven, and if they turn back to God, he places godly authority over them in the first heaven. This pattern is witnessed even in this dimension, repeatedly in the Old Testament, in God's dealings with the nation of Israel. When they were faithful to God, they were allowed to govern themselves, yet when they disobeyed repeatedly over time, God allowed them to be conquered (though not entirely, in the case of Israel). The only element I'm adding to this picture is that I'm showing how the establishment of authority over nations extends into the first heaven.

While many of the beings in the first heaven are faithful to God, most of them probably aren't. Earth in our current dimension is a mirror image of the first heaven, so all we have to do is look at the nations of the world today, in order to obtain a picture of what kind of beings are ruling in authority in the first heaven above said nations.

Government in the first heaven actually extends through the churches of that realm, which means that churches in our current dimension have more of an effect in government than most people think (Revelation 1:19–20). As for Satan, he currently rules the majority of the first heaven, though his power is not all encompassing. The archangel Michael, whom I previously mentioned, stands in defense of the territory over Israel because he's stated to be the prophet Daniel's prince, and Daniel was a prophet of Israel (Daniel 10:21).

> Revelation 1:19–20*
>
> Write the things which thou hast seen, and the things which are, and the things which shall be hereafter; The mystery of the seven stars which thou sawest in my right hand, and the seven golden candlesticks. The seven stars are the angels of the seven churches: and the seven candlesticks which thou sawest are the seven churches.

Following this scripture, chapters two and three of Revelation are letters addressed to angels ruling over churches on Earth. This indicates that God's governmental structure flows through angelic beings working through churches. Churches on Earth, therefore, effect government both in practical and supernatural ways.

Many other planets with life are most likely in a similar situation: divided into multiple dimensions, with each dimension divided into various ruling governmental structures. These higher realms, however, are very real, tangible realms, just as the Earth we live on. They are simply "phase-shifted"—existing within a different frequency, so to speak. Hebrews 1:1–2; 11:3 can't be any clearer concerning the fact that the universe consists of many *worlds*.

Hebrews 1:1–2 (bold emphasis added)

God, who at sundry times and in divers manners spake in time past unto the fathers by the prophets, Hath in these last days spoken unto us by his Son, whom he hath appointed heir of all things, **by whom also he made the worlds…**

Hebrews 11:3* (bold emphasis added)

Through faith we understand that **the worlds were framed by the word of God**, so that things which are seen were not made of things which do appear.

Aside from explaining that the universe consists of many worlds, Hebrews 11:3 is another example of proof that the Bible is divinely inspired. Atoms weren't discovered until 1803, yet the apostle Paul clearly states with knowledge rather than theory, that God made the universe we see using things that we can't see.[7]

The first war in the heavens (second heaven) began long ago, when Lucifer led an attack from Earth and into the highest heaven to take over, but failed. His kingdom was destroyed (Ezekiel 28:11–19; Isaiah 14:12–16). This war began thousands, millions, or perhaps even billions of years ago, before Adam and Eve were ever created. Technically speaking, this rebellion never ended.

Following the destruction of his kingdom, Satan continued spreading his deceptions and rebellion against God as he traversed the multidimensional universe of the heavens (second heaven) (Revelation 12:4). He continues to use limited access to the highest heaven to accuse the saints before God day and night. This limited access to the highest heaven will eventually be completely cut off, however (Revelation 12:9–10).

Revelation 12:9–10 (bold emphasis added)

And the great dragon was cast out, that old serpent, called the Devil, and Satan, which deceives the whole world: he was cast out into the earth, and his angels were cast out with him. And I heard a loud voice saying in heaven, Now is come salvation, and strength, and the kingdom of our God, and the power of his Christ: for the accuser of our brethren is cast down, **which accused them before our God day and night.**

During the war mentioned above, Satan and his band of fallen angels will eventually be completely forced out of the higher realms of glory. This is also mentioned in Ezekiel 28:16—Satan is restricted from the "stones of fire" (solar systems).

Ezekiel 28:16 (bold emphasis added)

By the multitude of thy merchandise they have filled the midst of thee with violence, and thou hast sinned: therefore I will cast thee as profane out of the mountain of God: and I will destroy thee, O covering cherub, **from the midst of the stones of fire.**

They will all end up confined to Earth in the days of the Antichrist (Isaiah 24:21–23; Revelation 12:4). When this absolute confinement occurs, it's likely that Satan will appear visibly to the people of Earth, in all manner of deceptive pageantry (2 Corinthians 11:13–14).

Isaiah 24:21

And it shall come to pass in that day, that the LORD shall punish the host of the high ones that are on high, and the kings of the earth upon the earth. And they shall be gathered together, as prisoners are gathered in the pit, and shall be shut up in the prison, and after many days shall they be visited. Then the moon shall be confounded, and the sun ashamed, when the LORD of hosts shall reign in Mount Zion, and in Jerusalem, and before his ancients gloriously.

Revelation 12:4

And his tail drew the third part of the stars of heaven, and did cast them to the earth: and the dragon stood before the woman who was ready to be delivered, for to devour her child as soon as it was born.

<u>2 Corinthians 11:13</u>

For such are false apostles, deceitful workers, transforming themselves into the apostles of Christ. And no marvel; for Satan himself is transformed into an angel of light.

This "strange god" of the Antichrist (Daniel 11:39) will arrive from among the stars and will cause everyone on Earth to marvel over him. He will be the primary otherworldly ally of the Antichrist, in addition to the fallen angel who will be released from the bottomless pit. More will be discussed about this later.

<u>Daniel 11:39</u> (bracketed comments added)

Thus shall he [the Antichrist] do in the strongest holds with a strange god, whom he shall acknowledge and increase with glory: and he shall cause them to rule over many, and shall divide the land for gain.

Interestingly, humans are, in the grand scheme of things, new kids on the block, but their role in the ancient battle of good versus evil is pivotal. It is through the human species that God has established his ultimate plan of grace, which will finally put an end to the rebellion against God. This is the primary reason angels are interested in the affairs of planet Earth. Most people live their lives from day to day, not realizing how big everything is. We of Earth are on the verge of the conclusion of an ancient war amid countless beings of unfathomable power—a war that has spanned across the eons, across innumerable galaxies, across dimensions; and it's all coming to an end on this little planet!

3. The Onion Theory

How is it that Earth can contain multiple realms, both inside of it and above it? Does it make sense that cohesive realms with living inhabitants conducting various activities can operate inside solid rock or molten lava, or that other realms containing various governmental structures and vast numbers of inhabitants can operate in the void of a cloudy sky? Furthermore, how can these realms coexist? I've already briefly mentioned some basic information about this theory, but now I'm going to go into more detail about it.

According to the apostle Paul in Romans 1:20, the invisible things in heaven are like the visible things on Earth. Paul, speaking of the third heaven (though the first heaven and other glorified worlds in the second heaven can be given similar descriptions), describes a glorified planet where God's throne is. It is a real, physical place, with natural terrain features, such as mountains (Revelation 21:10; Hebrews 12:22–23; Zechariah 6:1; Micah 4:1), rivers (Revelation 22:1–3), fountains of water (Revelation 7:17), and trees (Revelation 7:9; 22:1–3). There is a sky, just as on Earth, because of the presence of clouds, thunder, lightning, and hail (Revelation 3–8; 10:1–7; 11:19; take note of 11:19 in particular, which depicts thunder, lightning, and hail, all of which are weather conditions requiring an atmosphere. These were all witnessed by John while in the third heaven).

> Revelation 21:10 (bold emphasis added)
>
> And he carried me away in the spirit to **a great and high mountain**, and shewed me that **great city, the holy Jerusalem**, descending out of heaven from God...
>
> Hebrews 12:22–24* (bold emphasis added)
>
> But ye are come unto, and unto the city of the living God, the **heavenly Jerusalem**, and to an innumerable company of angels, to the general assembly and church of the firstborn, which are written in heaven, and to God the Judge of all, and to the spirits of just men made perfect...
>
> *The heavenly Jerusalem is in heaven, obviously, and is located on top of a mountain called Mount Zion.*
>
> Revelation 22:1–3* (bold emphasis added)
>
> And he shewed me a pure **river of water** of life, clear as crystal, proceeding out of the throne of God and of the Lamb. In the midst of the **street** of it, and on either side of the river, was there the **tree of life**, which bare twelve manner of fruits, and yielded her fruit every month: and the leaves of the tree were for the healing of the nations. And there shall be no more curse: but the throne of God and of the Lamb shall be in it; and his servants shall serve him...
>
> *While the description of New Jerusalem given here is in the future tense, this physical city with streets, rivers, and fruit-bearing trees with edible leaves currently exists, and is located in the highest heaven.*
>
> Revelation 11:19* (bold emphasis added)
>
> And the temple of God was opened in heaven, and there was seen in his temple the ark of his testament: and there were **lightning**, and voices, and **thunder**, and an **earthquake**, and great **hail**.

> *Aside from the elements of nature—lightning, thunder, and hail—other clearly physical objects are also mentioned, such as a temple, and "the ark of His testament." The event of an earthquake also reveals that this place is a planet, because earthquakes only occur on planets.*

In addition to the descriptions of nature in heaven, cities (Revelation 21), mansions (John 14:1–3), and streets (Revelation 12:21) are also present.

Revelation 21:2 (bold emphasis added)

And I John saw the holy **city**, New Jerusalem, coming down from God out of heaven, prepared as a bride adorned for her husband.

John 14:1–3 (bold emphasis added)

Let not your heart be troubled: ye believe in God, believe also in me. In my Father's house are many **mansions**: if it were not so, I would have told you. I go to prepare a place for you. And if I go and prepare a place for you, I will come again, and receive you unto myself; that where I am, there ye may be also.

As for the inhabitants of the highest heaven, animals such as horses (Revelation 19:11–14; Zechariah 1:8–11), wolves, lambs, lions, cattle, and oxen (Isaiah 11:6–8, 65:25) live there. People and angelic beings live there as well.

Zechariah 1:8

I saw by night, and behold a man riding upon a red horse, and he stood among the myrtle trees that were in the bottom; and behind him were there red horses, speckled, and white. Then said I, O my lord, what are these? And the angel that talked with me said unto me, I will show thee what these are. And the man that stood among the myrtle trees answered and said, These are they whom the LORD hath sent to walk to and fro through the earth. And they answered the angel of the LORD that stood among the myrtle trees, and said, We have walked to and fro through the earth, and, behold, all the earth sits still, and is at rest.

Revelation 19:11–14

And I saw heaven opened, and behold a white horse; and He that sat upon him was called Faithful and True, and in righteousness He doth judge and make war. His eyes were as a flame of fire, and on his head were many crowns; and he had a name written, that no man knew, but him himself. And He was clothed with a vesture dipped in blood: and his name is called The Word of

God. And the armies which were in heaven followed him upon white horses, clothed in fine linen, white and clean.

Isaiah 11:6–8

The wolf also shall dwell with the lamb, and the leopard shall lie down with the kid; and the calf and the young lion and the fatling together; and a little child shall lead them. And the cow and the bear shall feed; their young ones shall lie down together: and the lion shall eat straw like the ox. And the sucking child shall play on the hole of the asp, and the weaned child shall put his hand on the cockatrice' den.

Isaiah 65:25*

The wolf and the lamb shall feed together, and the lion shall eat straw like the bullock: and dust shall be the serpent's meat. They shall not hurt nor destroy in all my holy mountain, says the LORD.

Both references to animals in Isaiah are of the kingdom of God on Earth in the future, but because horses are clearly present in New Jerusalem in the third heaven now as depicted in Zechariah and Revelation, it can be derived that other animals are currently present in the third heaven as well.

What's more, a vast collection of miscellaneous items are mentioned, such as food, furniture, altars, fire and coals, incense and smoke, musical instruments, clothes, stones, books, bowls, crowns, lamps, doors and posts, girdles, banquets, walls, gates, precious jewels—the list goes on. Simply look up all references to visions of heaven, and these, as well as many other real, tangible things, will be described. Real, physical people have even been taken there, such as Enoch (Genesis 5:24) and Elijah (2 Kings 2:1). Some of the things in heaven are described as being somewhat beyond the glory of Earth, such as a "sea of glass" (Revelation 4:6, 15:2), "brass mountains" (Zechariah 6:1), "transparent gold," and pearls of immense proportions (Revelation 21:21), among others. Most of the things mentioned, however, can clearly be understood by things we already know.

Revelation 4:6

And before the throne there was a sea of glass like unto crystal: and in the midst of the throne, and round about the throne, were four beasts full of eyes before and behind.

Revelation 21:21

And the twelve gates were twelve pearls; every several gate was of one pearl: and the street of the city was pure gold, as it were transparent glass.

How could this place not be a glorified planet? Since the invisible things of heaven are like the visible things on Earth, then it isn't much of a stretch to conclude that the invisible things in and surrounding Earth are also like the visible things on Earth. Earth's coexisting realms are exactly like the Earth we all know and understand, with the exception that there are extremely awful places and extremely beautiful places of higher glory.

To make an analogy of how multiple realms can coexist in a layered fashion, and how their geographic locations can be the inner Earth, and the upper atmosphere of Earth, relative to our current dimension, consider the concept of frequency. A television is capable of receiving a bunch of different channels, all with the same cable. Each channel operates at a slightly different frequency than the other channels. In a similar fashion, Earth contains multiple levels of existence that appear much the same as the Earth we know, containing land, water, a sky, terrain features, and so on. As for the geographic correlation of these regions, the lower geographic regions contain the worse terrain (a great deal of fire, no water, extreme darkness), and the higher regions contain more glorious terrain. The full spectrum of the Earth, therefore, can be construed as a kind of onion, full of various layers of lithospheres (solid portion of the Earth). The lowest, inner layers contain the more degenerative degrees of the curse that God has cast upon Earth, and the upper regions in the sky still contain more glory than the Earth we currently live on. Being in the middle layer, the Earth we currently live on is actually the sky of the lower levels, and at the same time, it's the inside of the Earth of the first heaven.

Many people think of heaven as a formless cloud where people will have a kind of blind, blissful awareness. Or they think of heaven as a vast plane with a ground and a sky, but their thinking doesn't go beyond this. It's also common for people to give up on pondering what heaven is like because they label it as incomprehensible—but scripture clearly states that this is not the case. Heaven is composed of three realms: the sky, outer space, and a glorified planet where God's throne is. Each of these locations contains places much like the visible Earth. There are numerous descriptions of all the unseen realms in scripture. Romans 1:20 explains that these realms can be understood with precision, using descriptions of the visible things on Earth, and there's nothing confusing about them at all. In fact, in today's age of so many technological wonders and creative science

24   Aliens and the Antichrist

fiction writers, we most likely can have a clearer picture of what heaven is really like more than any generation preceding us—without actually seeing it in person.

A great deal of science fiction actually comes from various theories postulated in the field of quantum physics. The concept of a multiverse, dubbed *string theory*, is taken seriously by many mainstream scientists.[8] David Deutsch, premier scientist and winner of the Paul Dirac Medal and Prize, describes the concept of a multiverse in depth in his book, *The Fabric of Reality: The Science of Parallel Universes and Its Implications*. (Deutsch also contains a link to "Seti@Home" on his Web site, with a blurb stating, "My contribution to the UFO debate," indicating his belief in the existence of extraterrestrial life.)[9] While I certainly don't agree with his ideas espousing evolution, his concept of a multiverse is biblically accurate.

To many, the field of quantum physics is nothing more than gibberish, but to the Department of Defense and Department of Energy, practical applications for such *gibberish* are at the heart of their latest research and development projects. On the Defense Tech Web site, a controversial paper was recently published that outlines information about a motor that will propel a craft through *another dimension* at incredible speeds![10] That's right—another dimension, and I read this on a respected science Web site. Professor Jochem Hauser, one of the scientists involved with this new technology, is currently coordinating with NASA and the Air Force over the matter. He hopes to build a test device within the next five years. The science and technology section of Scotsman.com states the following about this proposed hyperspace drive:

> The hypothetical device, which has been outlined in principle but is based on a controversial theory about the fabric of the universe, could potentially allow a spacecraft to travel to Mars in three hours and journey to a star 11 light years away in just 80 days, according to a report in today's New Scientist magazine.[11]

The Z machine, an X-ray generator of enormous power developed by Sandia National Laboratories in New Mexico (and funded by the Department of Energy), might be used to test some of the basic science behind the functionality of the proposed hyperspace drive.[12] As for the Z machine, it's somewhat of an enigma as well. It generates an X-ray that lasts only for billionths of a second, but equates to 210 trillion watts, which is estimated at eighty times the entire world's output of electricity.[13]

The multidimensional nature of the universe is a biblical reality, and science is getting to the point of discovering this reality with physical proof. I don't need scientific proof like this, however. I feel as if I've always known that the heavens are close enough to see with my own eyes. I live with the knowledge that my ability to see into the first heaven is but the mere tuning of a dial that any angel is capable of adjusting. Nevertheless, even without an alteration of glory, or special technology, everyone has the opportunity to look through the first heaven and at the second heaven directly, every single night, just by gazing at the night sky. Any one of the stars in the night sky may be a sun with inhabited, orbiting planets.

4. Cursed Ground and Holy Ground

I thought it fitting to discuss cursed ground and holy ground before departing the topic of coexisting realms because these are areas where the dividing membrane between realms is thin. Perception into other dimensions can be made clearer by viewing these places as possible gateways and analyzing their characteristics.

In numerous places in scripture, there are references to cursed ground (Genesis 3:17; Isaiah 13:19–21; Jeremiah 50–51, and many others) and holy ground (Exodus 3:5; John 5:4). Following is an explanation of what I have found concerning these references.

Genesis 3:17 (cursed ground)

And unto Adam He said, Because thou hast hearkened unto the voice of thy wife, and hast eaten of the tree, of which I commanded thee, saying, Thou shalt not eat of it: cursed is the ground for thy sake; in sorrow you will eat of it all the days of thy life.

Isaiah 13:19* (cursed ground)

And Babylon, the glory of kingdoms, the beauty of the Chaldees' Excellency, shall be as when God overthrew Sodom and Gomorrah. It shall never be inhabited, neither shall it be dwelt in from generation to generation: neither shall the Arabian pitch tent there; neither shall the shepherds make their fold there. But wild beasts of the desert shall lie there; and their houses shall be full of doleful creatures; and owls shall dwell there, and satyrs shall dance there.

*The regions where Babylon, as well as Sodom and Gomorrah, used to be are desert wastelands even to this day.

Exodus 3:5 (holy ground)

And He said, Draw not nigh hither: put off thy shoes from off thy feet, for the place whereon thou stand is holy ground.

John 5:4 (holy ground/water)

For an angel went down at a certain season into the pool, and troubled the water: whosoever then first after the troubling of the water stepped in was made whole of whatsoever disease he had.

4.1. Cursed Ground

The Bible clearly states that the ground where an unjust killing (murder) is committed becomes cursed (Numbers 35:33–34). How exactly can we explain this, without sounding cryptic and abstract? What are the effects of cursed ground?

Numbers 35:33–34

So ye shall not pollute the land wherein ye are: for blood it defiles the land: and the land cannot be cleansed of the blood that is shed therein, but by the blood of him that shed it. Defile not therefore the land which ye shall inhabit, wherein I dwell: for I the LORD dwell among the children of Israel.

Simply put, the dividing element that separates dimensional realms from each other is the level of glory in a geographical region. In essence, God's glory, in some way yet undiscovered, is a physical property of the universe. It applies to living beings and nonliving matter as well. It can be considered a measurement of a spiritual proximity to God. In areas where God's presence is noted, the ground is said to be holy. In contrast, in areas where extreme acts of evil are committed, the level of glory is reduced over a geographical region (God's Holy Spirit recedes).

It is a theory of mine that cursed ground results in a thinning of the barrier that divides our current realm from the next lower level. In certain situations, negative effects of a lower degree of glory may transpire in areas of cursed ground. Cursed ground could possibly inflict illness on people, animals, and even plants, trees, and vegetation. Consider the curse that was cast on the area of ancient Babylon. Scripture states that God made this area a desolate wasteland where nothing will ever grow, and to this day, that curse holds true (Jeremiah 51:62–64).

Jeremiah 51:62 (bold emphasis added)

Then you will say, O LORD, thou hast spoken against this place, to cut it off, that none shall remain in it, neither man nor beast, but that **it shall be desolate for ever**.

Curses are associated with all manner of horrible things, many of which are documented throughout scripture, such as the many curses God caused to fall on the Egyptians through Moses (Exodus 3–12), the curses that plagued the Philistines when they captured the ark of the covenant from Israel (1 Samuel 5–6), and the list goes on. These scriptures make it plausible to suggest that cursed ground may resonate with numerous negative effects, and influence people and animals in other negative ways, such as making them more violent, lustful, selfish, or sickly. Many of these negative effects may be due to people simply spending time in areas where the atmosphere has a lower level of glory. Such cursed places may also serve as portals, where beings from the lower realms might pass through into our current realm. Some of these beings are demons—creatures so far from God they are void of any kind of morality. In other cases, the beings in question might be the spirits of the deceased.

4.2. What Does the Bible Say About Ghosts?

While this book focuses mainly on aliens, I will briefly provide a few details on ghosts, simply because, like aliens, ghosts are otherworldly beings. For starters it's not typical of Christians to believe in the existence of ghosts (disincarnate human spirits inhabiting this world), but scripture gives several indications that ghosts are a real phenomenon, and that they are, in fact, exactly what people have been saying they are: the disincarnate spirits of people who have physically died. Consider the fact that Jesus himself never directly stated that ghosts do not exist. Instead he actually defined the nature of ghosts when he told those who saw him after his resurrection that he wasn't a ghost because he was physical, made of flesh and bone that people could touch and feel (Luke 24:36–39). If ghosts aren't real, then why didn't he clear up the matter and state directly, "I'm not a ghost because ghosts don't exist"?

> Luke 24:36–39 (This quote is from the International Standard Version of the Bible, because of the more accurate interpretational use of the word "ghost").
>
> While they were talking about this, Jesus himself stood among them and said to them, Peace be with you. They were startled and terrified, thinking they were seeing a ghost. He said to them, Why are you frightened, and why are doubts arising in your hearts? Look at my hands and my feet, for it is I myself. Touch me and see, for a ghost doesn't have flesh and bones as you see that I have.

In some versions of the Bible, the word "spirit" is substituted for "ghost" in the above passage, but this substitution can be a cause for confusion, at least in the English language. All ghosts are spirits, but not all spirits are ghosts.

In the passage above, Jesus defines ghosts as noncorporeal spirits who have no flesh whatsoever. Angels, on the other hand, are considered spirits as well, but they have physical bodies made of a spiritual flesh; they are capable of physical activities. The Bible gives many examples of angels exhibiting physical qualities that ghosts do not have. Confusing matters even more, angelic spirits who aren't ghosts can act as if they are ghosts if they want to (exhibit nonphysical qualities), such as walking through walls, flying, and so on. Ghosts, on the other hand, do not have the option of being physical. On rare occasion, poltergeists (a specific type of ghost) might have enough energy to move objects, but that's about all they can do in the physical realm without possessing a host.[14] If this explanation isn't good enough, then just keep reading, and hopefully things will become clearer later on. Or not...

Continuing on, scripture speaks of two types of ghosts; glorified disincarnate entities, and cursed disincarnate entities.

4.2.1. Glorified Disincarnate Entities

Since the resurrection of Christ, the apostle Paul stated that for Christians, to be apart from the body is to be in the presence of the Lord (2 Corinthians 5:8). Many take this to mean that these spirits never return to Earth until they are reunified with their physical bodies, but scripture never directly states that it is impossible for said spirits to travel anywhere. Instead the book of Revelation refers to one of the angels of John's vision as being a man (Revelation 21:17, 22:9), and further defines him as one of the brethren. Noting that glorified saints become angels, I ask the following questions: Are angels only in God's presence and never allowed to go anywhere else? Is there no such thing as an angelic assignment? Aren't angels considered messengers?

> Revelation 21:17*
>
> And he measured the wall thereof, an hundred and forty and four cubits, according to the measure of a man, that is, of the angel.
>
> *John says that the angel he's talking with in his vision is a man. This man later identifies himself as being a "fellow servant," "brother," and "prophet" in Revelation 22:9, shown below. This would not be true of an angel who was never a human on Earth.

Revelation 22:9

Then said he unto me, See thou do it not: for I am thy fellow servant, and of thy brethren the prophets and of them which keep the sayings of this book: worship God.

Further illustrating my point, how is it that Moses and Elijah met with Jesus on the mount of transfiguration, when in the book of Deuteronomy, chapter 34, it states that Moses died, and God buried him in the desert? Moses was a glorified ghost, visiting Jesus from heaven. Even Abel, the first naturally born human being to walk the Earth, became a haunting ghost after being murdered by his brother. The Bible specifically states that his blood cried from the soil, and God heard it (Genesis 4:10–11). Then in the New Testament, the apostle Paul states that Abel achieved his ability to reach out beyond the grave "by faith" (Hebrews 11:4).

Genesis 4:10–11 (bold emphasis added)

And he said, What hast thou done? **The voice of thy brother's blood cries unto me from the ground**. And now art thou cursed from the earth, which hath opened her mouth to receive thy brother's blood from thy hand...

Hebrews 11:4 (bold emphasis added)

By faith, Abel offered unto God a more excellent sacrifice than Cain, by which he obtained witness that he was righteous, God testifying of his gifts: **and by it, he being dead yet speaks**.

Finally, a third example of a good ghost is found in 1 Samuel 28:11–15, when the ghost of the prophet Samuel was conjured by a medium at the request of King Saul. The experience was not a pleasant one for King Saul.

1 Samuel 28:11

Then said the woman, Whom shall I bring up unto thee? And he said, Bring me up Samuel. And when the woman saw Samuel, she cried with a loud voice: and the woman spoke to Saul, saying, Why hast thou deceived me? For thou art Saul. And the king said unto her, Be not afraid: for what saw thou? And the woman said unto Saul, I saw gods ascending out of the earth. And he said unto her, What form is he of? And she said, An old man cometh up; and he is covered with a mantle. And Saul perceived that it was Samuel, and he stooped with his face to the ground, and bowed himself. And Samuel said to Saul, Why hast thou disquieted me, to bring me up? And Saul answered, I am sore distressed; for the Philistines make war against me, and God is departed from me, and answers me no more, neither by prophets, nor by dreams: therefore I have called thee, that thou might make known unto me what I shall do.

Concerning glorified ghosts, my great-grandmother saw her husband in the spirit standing at the foot of her bed the day before she died. She was a godly woman, and he was a godly man when he was alive, fully devoted to God—a traveling evangelist, in fact. My Grandma Thelma told her daughter that he was young, and glowing in the spirit, surrounded by light. It was an incredibly beautiful experience as she recalled it. He was holding his hands out to her, calling her to come to him. The next day, she died.

4.2.2. Cursed Disincarnate Entities

Ghostly encounters with glorified disincarnate entities aren't nearly as common as ghostly encounters with cursed disincarnate entities. Cursed disincarnate entities are the spirits of those who are not saved. Most Christians assume this means that they automatically go to hell, therefore they are no longer in the realm of the living in any way. I think this is only partially true, however. It is true that these beings go to a place the Bible calls hell…but where is hell? Most people know that the Bible speaks of hell as being inside Earth, and most people assume it's in another dimension too far away to perceive. What most people don't consider, however, is the possibility that in certain places where the level of glory is lower on the surface of Earth (cursed ground), people might be able to perceive some of the *upper dimensions of hell*, because hell contains various levels of existence, some of which are close to our own realm. If this isn't the case, then how are demons able to enter our realm and possess people? If demons can possess people, then why can't the spirits of deceased people enter our realm and possess people as well? After all, a demon is a ghost too (nonhuman disincarnate spirits of the Nephilim and the pre-Adamite civilization, which I will discuss later). Why would the disincarnate spirits of people have stricter limitations than beings who are even more evil than they are?

I've heard the argument that disincarnate human spirits can't leave hell because in the parable of Lazarus and the rich man, the rich man requested that Lazarus warn his (the rich man's) relatives about hell, so they would repent. Lazarus never left Abraham's bosom (located in the paradise portion of Hades), however, even to warn the rich man's living family members to repent (Luke 16:19–31).

Luke 16:19–31 (bold emphasis added)

There was a certain rich man, which was clothed in purple and fine linen, and fared sumptuously every day: And there was a certain beggar named Lazarus, which was laid at his gate, full of sores, And desiring to be fed with the crumbs which fell from the rich man's table: moreover the dogs came and licked his sores. And it came to pass, that the beggar died, and was carried by the angels into Abraham's bosom: the rich man also died, and was buried; And in hell he lift up his eyes, being in torments, and seeth Abraham afar off, and Lazarus in his bosom. And he cried and said, Father Abraham, have mercy on me, and send Lazarus, that he may dip the tip of his finger in water, and cool my tongue; for I am tormented in this flame. But Abraham said, Son, remember that thou in thy lifetime receivedst thy good things, and likewise Lazarus evil things: but now he is comforted, and thou art tormented. And beside all this, between us and you there is a great gulf fixed: so that they which would pass from hence to you cannot; neither can they pass to us, that would come from thence. **Then he said, I pray thee therefore, father, that thou wouldest send him to my father's house: For I have five brethren; that he may testify unto them, lest they also come into this place of torment. Abraham saith unto him, They have Moses and the prophets; let them hear them.** And he said, Nay, father Abraham: but if one went unto them from the dead, they will repent. And he said unto him, If they hear not Moses and the prophets, neither will they be persuaded, though one rose from the dead.

Because of this, it is commonly assumed by most Christians that no one would ever be allowed to leave hell (even the paradise portion of hell, which no longer exists) for any reason. I see a valid point in this reasoning, because I've never heard of anyone being warned by a deceased relative to repent or burn in hell. But I still don't think this one parable about this one person is sufficient to draw a conclusion that applies to every situation. This one parable does not definitively disprove the existence of ghosts. The simple fact remains that the Bible documents many demonic possessions, and there has to be an explanation as to how this takes place.

What if, for example, a murder curses the ground and creates an atmospheric anomaly that allows specific cursed beings access into that region of space? After all, Satan and his minions only have power in places where they're given a foothold. This rule probably applies to people too. The rich man clearly had no right or authority to enter into Abraham's bosom, nor did he have any right or authority to be given special access so he could warn his family members who were still alive—but then again, the rich man wasn't brutally murdered. If he were, he might have been given special access to the location of his murder. Having conducted a little research of my own into the phenomenon of haunting, I have found that premature death, especially in cases involving murder, is a key factor in what creates haunting phenomenon. According to a wide variety of sources referenced in *Harper's Encyclopedia of Mystical & Paranormal Experience*, "haunted sites seem to be places merely frequented or liked by the deceased, or places where violent death has occurred."[15]

Scripture speaks a great deal about the importance of keeping a land spiritually clean and free of sin, and I suspect this has something to do with spiritually cursed, disincarnate entities gaining access into the realm of the living, at least to some limited extent. Once they have limited access, they test their boundaries, the same as any criminal would in a situation where freedom might be obtainable. How exactly do people get possessed, anyway? Clearly when the conditions are right (for example, when a person who is lost, spiritually weak, and living in sin, and on top of all of that goes to places like palm-reading shops, cemeteries, satanic churches, or haunted houses—which are most likely cursed ground), demonic possession becomes possible. If this were not the case, it would never occur. Certainly something has to happen. Because demonic possession is possible, and it entails a cursed disincarnate entity managing to escape its confines in hell, then I see no reason why disincarnate human spirits can't do the same thing. Dr. Edith Fiore, author of *The Unquiet Dead: A Psychologist Treats Spirit Possession*, seems to think so as well. While Dr. Fiore's work relies heavily on hypnotherapy, and is tarnished with her beliefs in reincarnation, her findings, which document numerous cases of people being possessed by spirits of the deceased, are certainly worthy of further investigation.

While many people believe that all ghostly activity is solely the work of demons, I don't think the Bible supports this view. In addition to what I've already stated, the Bible provides an array of terms used to describe possession, some of which may refer to spirits that are not necessarily demons. Terms such as "evil spirits" (Judges 9:23; 1 Samuel 16:14–16, 16:23, 18:10, 19:9; Acts 19:12–13, 19:15–16; Luke 7:21, 8:2), "unclean spirits" (Matthew 10:1; Mark 1:27, 3:11, 5:13, 6:7; Luke 4:36, 6:18; Acts 5:16, 8:7; Revelation 16:13), "foul spirits" (Mark 9:25; Revelation 18:2), and "familiar spirits" (Leviticus 19:31, 20:6; Deuteronomy 18:11; 1 Samuel 28:3, 28:9; 2 Kings 21:6, 23:24; Isaiah 8:19, 19:3) might refer to the deceased spirits of people, rather than demons.

In any case, being possessed is not healthy, whether it's by a demon or a deceased person. Furthermore, in today's New Age climate, being possessed by a disincarnate person may deceive people into believing false doctrines like reincarnation, which according to Hebrews 9:27 can't be true. A person possessed with the disincarnate spirits of other people may misinterpret the memories of the spirits that possess them as their own memories and think that they are remembering their own past lives; but these people aren't reincarnated, they're possessed! This was exactly the case with many of Dr. Fiore's patients. These people needed deliverance and healing, rather than hypnotherapy sessions to investigate their past lives with enthusiasm, as Dr. Fiore and other psychologists, such as Dr. Bruce Goldberg, have been known to do.

Hebrews 9:27

And as it is appointed unto men once to die, but after this the judgment...

4.2.3. Interaction with Cursed Disincarnate Entities

Concerning the ghosts of cursed disincarnate entities, one fact is made abundantly clear in scripture: people are forbidden to seek them out to interact with them (Leviticus 19:26; Deuteronomy 18:9–15; 2 Kings 17:17–18; 2 Chronicles 33:6).

Leviticus 19:26

Ye shall not eat anything with the blood: neither shall ye use enchantment, nor observe times.

Deuteronomy 18:9–15

When thou art come into the land which the LORD thy God gives thee, thou shalt not learn to do after the abominations of those nations. There shall not be found among you any one that makes his son or his daughter to pass through the fire, or that uses divination, or an observer of times, or an enchanter, or a witch, or a charmer, or a consulter with familiar spirits, or a wizard, or a necromancer. For all that do these things are an abomination unto the LORD: and because of these abominations the LORD thy God doth drive them out from before thee. Thou shalt be perfect with the LORD thy God. For these nations, which thou shalt possess, hearkened unto observers of times, and unto diviners: but as for thee, the LORD thy God hath not suffered thee so to do. The LORD thy God will raise up unto thee a Prophet from the midst of thee, of thy brethren, like unto me; unto him ye shall hearken...

Nowadays mediums even have their own TV shows, where members of an audience are given an opportunity to communicate with the deceased.[16] These mediums will speak about God and spread their teachings and philosophies of the spiritual realm, as if they are spiritually enlightened, yet they make their living by glorifying their sinful activities on national television. They obviously don't think they're sinning, but it should be noted that they don't take their definition of sin from the Bible.

Just about everyone has a few ghost stories, and I don't see how it could be a sin if someone just happens to see a ghost. Some people see and hear them all the time, however (I actually know a few people who have this ability), and here's where a distinction must be made. People who have this ability should use it for what God wants them to use it for, rather than blindly wandering about in the dangers of the spiritual realm, believing the words of every voice that speaks. All true mediums probably start out with a God-given spiritual gift, but it becomes corrupted after years of subjection to deception.

Interaction with the deceased should only be taken to the point of testing the spirits (1 John 4:1–4), but once the test results are in, if the spirits in question don't pass the tests, all communication should immediately cease. Interaction with the deceased is dangerous because the lower realms are full of evil spirits seeking ways to penetrate the physical realm for the sole purpose of pursuing their own agendas, which vary widely but never include the will of God.

> 1 John 4:1–4
>
> Beloved, believe not every spirit, but try the spirits whether they are of God: because many false prophets are gone out into the world. Hereby know ye the Spirit of God: Every spirit that confesses that Jesus Christ is come in the flesh is of God: And every spirit that confesses not that Jesus Christ is come in the flesh is not of God: and this is that spirit of Antichrist, whereof ye have heard that it should come; and even now already is it in the world.

Interaction with *spirits* isn't completely forbidden, because if it were, the prophets Daniel, Ezekiel, the apostle John, and Jesus himself would be guilty of sin, but this is clearly not the case. John spoke with the deceased spirit of a prophet (Revelation 22:9), Daniel and Ezekiel spoke with beings (angels) who could've been construed as ghosts, and Jesus spoke with Moses after Moses was physically deceased (Matthew 17:3; Mark 9:4). The difference in all of these cases, however, is that these spirits *passed the test*. They were *good spirits*, faithful to God, not evil spirits.

The reason for the prohibition of interaction with deceased spirits should be obvious. The barrier that divides dimensions from each other is there to keep the corruption of lower realms from spreading into the higher realms. If beings are stuck in the lower realms, it's because they are cursed in great measure. Their appearances, questions, activities, and requests can never fully be trusted, and that is why they must be tested if encountered (1 John 4:1–3). Personally I would include even more tests over and above 1 John 4:1–4 because spirits can be very tricky! (I know—and don't ask how.) In short what they say should line up with all of the truths outlined in the Bible. If they say something that doesn't quite fit—then bingo! They need to take a hike.

A demon-possessed man can serve prominently in ministry (Judas Iscariot did), and a true believing man of God might murder an innocent man in order to ensure a marriage to that man's wife (King David did), so the human capacity to judge good and evil is hampered already without adding the element of being disincarnate. It is possible that disincarnate spirits may want simple things, and some of them may even do good things, just as non-Christian people who are living may want simple things and even do good things—but the bottom line is that cursed, disincarnate spirits are not in heaven for a reason. This fact cannot be ignored, and neither can the warnings clearly outlined in scripture.

4.2.4. Advice for People Who See and Hear Spirits of the Deceased

For anyone with the spiritual gift of being able to see and hear spirits, this gift may be a powerful gift from God, with a God-given purpose. Simply reviewing all the spiritual gifts listed in scripture can give a clue as to what this ability might entail (1 Corinthians 12). For example, if someone sees or hears deceased spirits, then that should be a cue to pray with specific knowledge over people and places. In other words, if you see a demon possessing or manipulating someone, cast it out in the name of Jesus! That's what Jesus did. He had the ability to detect spirits because he knew when Satan and demons were hanging around, manipulating people (Matthew 16:23; Mark 8:33; Luke 4:8). He never wasted his time in casual conversation with such entities either. He cast them out and rebuked them left and right (Matthew 8:16, 8:31; Mark 1:34, 6:13, 16:9; Luke 13:32, and many more). We should do the same. I imagine such a gift would be particularly handy for keeping one's eyes on the true enemy, rather than on all the pawns. One person I know personally who has this ability serves in deliverance ministry in my church, and conducts spiritual warfare of an unusual nature; the results are remarkable.

The ability to channel spirits is also a God-given gift, though it is intended *only* for channeling the power of the Holy Spirit, and none other. The satanic perversion of this is doing exactly the opposite. In today's New Age climate, apparently anyone can learn how to channel spirits by following instructions in a number of widely available books. Mediums are magnets for demonic possession—probably more susceptible to it than people without such gifts are. Demons seek out people who can perceive them because they can exude a greater influence through these people who are natural conduits. In essence the people whom God intends to use to free people from satanic oppression and possession can sometimes become slaves to Satan when they fall into the seduction of glorifying themselves and obtaining profit and prosperity by doing things with their spiritual gifts that are forbidden in scripture. The temptation to misuse powers is a constant temptation for many psychics, according to an article found in *The Journal of the American Society for Psychical Research*.[17]

There is a prophet mentioned in the Old Testament who had powerful spiritual abilities, and he was tempted to do things with his spiritual gifts that were against the will of God (Numbers 22–24; Nehemiah 13:2). An interesting thing to note about Balaam is that he prayed and spoke with God, and God used him to bless the nation of Israel, which proves his gifts were originally from God and intended to bring glory to God. To suggest otherwise would be suggesting that God used a man with satanic power to bless the nation of Israel. Despite the fact that Balaam's gifts were from God, Balaam ended up using them to serve his own purposes. His God-given gifts eventually became satanic because of this (Revelation 2:14; Jude 1:11; 2 Peter 2:15) and brought about his own premature death (Numbers 31:8).

> Numbers 22:2, 5–12
>
> And Balak the son of Zippor saw all that Israel had done to the Amorites. He sent messengers therefore unto Balaam the son of Beor to Pethor, which is by the river of the land of the children of his people, to call him, saying, Behold, there is a people come out from Egypt: behold, they cover the face of the earth, and they abide over against me: Come now therefore, I pray thee, curse me this people; for they are too mighty for me: peradventure I shall prevail, that we may smite them, and that I may drive them out of the land: for I know that he whom thou blesses is blessed, and he whom thou curses is cursed. And the elders of Moab and the elders of Midian departed with the rewards of divination in their hand; and they came unto Balaam, and spoke unto him the words of Balak. And he said unto them, Lodge here this night, and I will bring you word again, as the LORD shall speak unto me: and the princes of Moab abode with Balaam. And God came unto Balaam, and said, What men are these with thee? And Balaam said unto God, Balak the son of Zippor, king of Moab, hath sent unto me, saying, Behold, there is a people come out of Egypt, which cover the face of the earth: come now, curse me them; peradventure I shall be able to overcome them, and drive them out.

And God said unto Balaam, Thou shalt not go with them; thou shalt not curse the people: for they are blessed.

Revelation 2:14

But I have a few things against thee, because thou hast there them that hold the doctrine of Balaam, who taught Balak to cast a stumbling block before the children of Israel, to eat things sacrificed unto idols, and to commit fornication.

Jude 1:11

Woe unto them! For they have gone in the way of Cain, and ran greedily after the error of Balaam for reward, and perished in the gainsaying of Core.

2 Peter 2:15

Which have forsaken the right way, and are gone astray, following the way of Balaam the son of Bosor, who loved the wages of unrighteousness…

Numbers 31:8

And they slew the kings of Midian, beside the rest of them that were slain; namely, Evi, and Rekem, and Zur, and Hur, and Reba, five kings of Midian: Balaam also the son of Beor they slew with the sword.

4.2.5. Ghostly Visitations of Loved Ones

Some people reading this may be having a troubled feeling concerning a personal experience. Perhaps family members have had visitations of what they perceived as deceased loved ones. Keep in mind that this doesn't mean that the loved ones are most certainly in hell without hope. The visitation could have been like my great-grandmother's visitation. Another possibility could be that a demon, or even another disincarnate human who learned such a feat, could appear in any form he/she/it desired. There may be a particular benefit to appearing to someone in the form of a deceased loved one. In any case, such visitations can have multiple explanations, many of which don't conclude with the eternal damnation of loved ones.

4.2.6. Beacons of Injustice

The discussion of cursed ground started my query into the origin of deceased spirits. The Old Testament book of Leviticus states that the act of murder will curse the ground (Leviticus 17:10–15). The only way to remove such a curse is by the blood of the offender. In other words, curses require justice in order to be removed. When this is considered, it is by no coincidence that haunting activity frequently goes hand in hand with brutal murders.[18] In this respect, the haunting serves as a beacon of injustice.

Does this mean that all haunted locations are inhabited with actual sentient beings? Are all of those beings truly ghosts, or could there be something else? Conducting a little research into this matter, I've found something interesting worth mentioning while on the subject of ghosts. Concerning some haunted locations in particular, the activity in question sometimes resembles a recording more than it does a sentient being, or ghost. In some cases, for example, a haunting may be the reenactment of an event, such as a war (like the well-famed Battle of Gettysburg haunting), or a crime, rather than a particular being who seems lost and confused.

Concerning the Gettysburg haunting, an article on hauntedhouses.com, which I have quoted below, documents an amazing incident that happened during the filming of the 1993 film, *Gettysburg*.

> During the filming of the film, *Gettysburg*, actors/Civil War re-enactors dressed in uniform would often walk off set, exploring the real battlefields during their down time. A group of Union soldiers went up to the Little Round Top area, to enjoy the lovely sunset. While standing there, they heard the rustling of leaves behind them. Looking to see who it was, imagine their surprise when a rather haggard looking old man, dressed as a Union private, made his appearance. The man was filthy and smelled of sulfur, a key ingredient of the black powder used in 1863. He walked up to the men and handed them a few musket rounds. He said, "Rough one today, eh boys?" He then turned and walked away. As the startled men examined the musket rounds, this mystery man vanished into thin air! They brought these musket rounds into the town of Gettysburg, and they were authenticated as original rounds, 130 years old![19]

Was this a real person that had died, or was something else happening here? Could this have been a distortion in the space-time continuum, allowing physical matter to pass 130 years into the future?

Haunting activity can also include inanimate objects, such as a ghost train (the purported train of President Lincoln's funeral),[20] a ghostly wagon from the cowboy days,[21] or the sound of musical instruments being played.[22] What could this strange energy be, if it was never alive to begin with?

When I wrote *Aliens in the Bible*, I went into detail discussing the concept of thought-forms.[23] In short, thought-forms are the energy of thoughts that are—in some way yet to be discovered—infused into matter. (Read *Aliens in the Bible* for more reference material concerning scientific and scriptural support for the existence of thought-forms.)

What inspired my thinking about how thought-forms and haunting activity might be related was a 1992 movie titled *To Catch a Killer*.[24] This movie chronicled the rampages of a serial killer. What caught my attention about it was that a woman with the psychic ability of psychometry helped to apprehend a serial killer. When handed a camera and a high-school ring, the woman was able to describe the killer's physical appearance and personality characteristics. She was able to tell that the owner of the objects was dead, and knew a number of details about the location of his dead body. Her abilities were almost like a bloodhound's ability to identify with a scent. An unknown, undetectable energy surrounded the objects in question, and they were chock-full of information. While the energy itself wasn't alive, the information it held was audio and visual, contained smells, and even conveyed emotions. Concerning this energy, it could very well be that God ordains its existence, so people with certain spiritual gifts can tap into it and obtain information.

In summation concerning the existence of ghosts, I believe that the universe is full of things that can't be easily identified. Ghosts could be hallucinations, demonic impersonators, fluctuations in the time-space continuum, the actual spirits of the deceased, or free-roaming energy that contains many attributes of spiritual beings, but isn't actually alive. A number of odd combinations of these could be at work as well. Cursed ground might be teeming with negative energy, which in turn attracts demonic spirits. Perhaps demons are able to tap into free-roaming energy sources. The possibilities could be endless.

The cure for haunting is the same now as it was in the Old Testament: if justice is served concerning a crime that caused a particular curse, then the curse will be removed from the ground, and the haunting will end. Certain situations might arise, however, where it may not be humanly possible for justice to be served. For example, some documented cases of haunting are several hundred years old. Anyone involved in said criminal activity would long be dead, without any perpetrator ever coming to justice. What is to be done in such places?

The blood of Jesus was shed for the sins of the world! Because of this, it is possible to remove curses by symbolically applying the blood of Jesus through prayer. Family members and church congregations should pray over haunted houses. Sometimes haunting activity can be persistent (such was the case with my own house), but people can be persistent right back! Some places may have generations of curses placed on them. I imagine some of the torture chambers in the ancient castles of Europe are like this, but the blood of Jesus is more than sufficient to cover these sins. Pray, pray, pray, and eventually the haunting will go away. (Prayer needs to go hand in hand with one's lifestyle as well. It does no good to pray without believing, and evidence of true faith should be obvious in one's lifestyle.)

4.3. Holy Ground

We know from scripture that committing heinously wicked acts creates cursed ground, so what exactly does it take to glorify a place and declare it holy ground? While curses are the product of a separation from God's presence, the close proximity of God's presence establishes holy ground.

When Moses encountered God, God told Moses to remove his sandals because he was standing on holy ground (Exodus 3:5; Acts 7:33). The ground was holy because God was in near proximity. If that exact location could be found today, I wouldn't be surprised if people would be cured of practically anything just by spending time there because it is a unique place on Earth where God manifested himself in a tangible form.

> Exodus 3:5
>
> And He said, Draw not nigh hither: put off thy shoes from off thy feet, for the place whereon thou stand is holy ground.
>
> Acts 7:33
>
> Then said the Lord to him, Put off thy shoes from thy feet: for the place where thou stand is holy ground.

Cursed ground has negative side effects; holy ground has glorious side effects. God's reason for making Moses remove his sandals might be construed as a form of respect, but I think it also included giving Moses a special blessing. Call it my absurd opinion, but I think Moses removing his sandals resulted in more of God's glory flowing through his body. Moses lived to be 120 years old (Deuteronomy 34:7), and would've lived longer, had he not made a horrible mistake and disobeyed God during his years of leading Israel (Deuteronomy 32:48–52). He probably wasn't sick a day in his life. Moses was also clearly anointed with God's power like few people who have ever walked the Earth. The reason for this is simple: he spent time in God's direct, manifest presence.

> Deuteronomy 32:48–52
>
> And the LORD spoke unto Moses that selfsame day, saying, Get thee up into this mountain Abarim, unto mount Nebo, which is in the land of Moab, that is over against Jericho; and behold the land of Canaan, which I give unto the children of Israel for a possession: And die in the mount whither thou go up, and be gathered unto thy people; as Aaron thy brother died in mount Hor, and was gathered unto his people: Because ye trespassed against me among the children of Israel at the waters of Meribah-Kadesh, in the wilderness of Zin; because ye sanctified me not in the midst of the children of Israel. Yet thou shalt see the land before thee; but thou shalt not go thither unto the land which I give the children of Israel.

Jesus, of course, created holy ground wherever he went. A woman who touched the hem of his garment was healed; even his clothing was full of God's glory. To this day, people have claimed healing from seeing the Shroud of Turin, the burial cloth that Jesus left behind in his tomb after he was resurrected.[25] While the Shroud of Turin is a relic of debated authenticity, if it is truly authentic, it probably does contain healing power, because the entire Earth increased in glory the second Jesus died. The power of God filled Earth to such a degree at that time that numerous people were even resurrected from their graves (Matthew 27:52–53).

> Matthew 27:52–53
>
> And the graves were opened; and many bodies of the saints, which slept, arose, and came out of the graves after his resurrection, and went into the holy city, and appeared unto many.

God's presence creates holy ground, and God's presence flowing through his servants can create holy ground as well. As the apostle Peter walked through the streets, his shadow healed people when it passed over them. As for the Old Testament prophet Elisha, even the bones of his dead body held enough spiritual residue of God's presence that they resurrected a man from the dead (2 Kings 13:21).

Acts 5:15

Insomuch that they brought forth the sick into the streets, and laid them on beds and couches, that at the least the shadow of Peter passing by might overshadow some of them.

2 Kings 13:21

And it came to pass, as they were burying a man, that, behold, they spied a band of men; and they cast the man into the sepulchre of Elisha: and when the man was let down, and touched the bones of Elisha, he revived, and stood up on his feet.

Angels create holy ground and holy water as well, which is the aquatic equivalent. The pool of Bethesda that Jesus visited during his ministry (John 5:2–4) was a place that contained holy water. Just as I theorized about cursed ground creating portals where beings of lower realms can penetrate our own realm, this place of holy ground seemed to be a portal where beings from a higher realm entered our current realm. Angels would periodically touch the waters of this pool. On such occasions, the first one to touch the water when it stirred would be healed.

John 5:2

Now there is at Jerusalem by the sheep market a pool, which is called in the Hebrew tongue Bethesda, having five porches. In these lay a great multitude of impotent folk, of blind, halt, withered, waiting for the moving of the water. For an angel went down at a certain season into the pool, and troubled the water: whosoever then first after the troubling of the water stepped in was made whole of whatsoever disease he had.

5. Interplanetary/Interdimensional Travel

I began by speaking about the heavens, then about holy and cursed ground and their relation to overlapping realms of existence in a multidimensional universe. I then fell into the tangent of ghosts when speaking of how beings from other dimensions may be perceived by people from our own dimension (or level of glory). I will now bring relevance to the tangent of ghosts, by backing away from Earth and redirecting my focus back into the heavens. The same concept of interdimensional beings (ghosts being one example) can be examined at the macro level.

Among the highly advanced beings who exist throughout the cosmos, some have achieved the ability to access other worlds or dimensions (whether through innate abilities, technological means, or other methods). Following are a few of the caveats of said activity.

5.1. Interplanetary/Interdimensional Travel Conducted by Cursed Beings

Scripture provides a number of examples of cursed beings who achieved or were innately capable of interplanetary/interdimensional travel. This access appears to have certain limitations, however. Concerning fallen angels, Satan once had unlimited access to the highest heaven, but since his fall from glory, his kingdom was destroyed (Isaiah 14:12–14; Ezekiel 28:11–19), and his access to the highest heaven has been limited.

<u>Isaiah 14:12–16</u> (bold emphasis added)

How art thou fallen from heaven, O Lucifer, son of the morning! **How art thou cut down to the ground,** which didst weaken the nations! For thou hast said in your heart, I will ascend into heaven, I will exalt my throne above the stars of God: I will sit also upon the mount of the congregation, in the sides of the north: I will ascend above the heights of the clouds; I will be like the most High. **Yet thou shalt be brought down to hell, to the sides of the pit.** They that see thee shall narrowly look upon thee, and consider thee, saying, Is this the man that made the earth to tremble, that did shake kingdoms?

Currently, Satan and his band of fallen angels are still allowed to attend divine counsels with God and his faithful angels, as described by the Old Testament prophet Job (Job 1:5–6; 2:1). For the most part, he attends in order to make accusations against the saints (Revelation 12:10).

<u>Job 1:5–6</u> (bold emphasis added)

And it was so, when the days of their feasting were gone about, that Job sent and sanctified them, and rose up early in the morning, and offered burnt offerings according to the number of them all: for Job said, It may be that my sons have sinned, and cursed God in their hearts. Thus did Job continually. Now there was a day when the sons of God came to present themselves before the LORD, and **Satan came also among them.**

<u>Revelation 12:10</u> (bold emphasis added)

And I heard a loud voice saying in heaven, Now is come salvation, and strength, and the kingdom of our God, and the power of his Christ: for **the**

accuser of our brethren is cast down, which accused them before our God day and night.

Concerning Satan's access to glorified realms in the second heaven, this too was (or is still in the process of being) eventually terminated. In Ezekiel 28:15–16, "stones of fire" most likely refers to solar systems (suns) in the heavens.

Ezekiel 28:15–16 (bold emphasis added)

Thou wast perfect in thy ways from the day that thou wast created, till iniquity was found in thee. By the multitude of thy merchandise they have filled the midst of thee with violence, and **thou hast sinned: therefore I will cast thee as profane out of the mountain of God: and I will destroy thee, O covering cherub, from the midst of the stones of fire**.

Satan's restriction from the "stones of fire" did not occur, however, before he spread his rebellious campaign across the universe (glorified realms in the second heaven). Scripture states that Satan deceived one-third of the inhabitants of the host of heaven: In Revelation 12:4, "the third part of the stars of heaven" that Satan deceived, refers to angels, which live among the stars.

Revelation 12:4 (bold emphasis added)

And his tail drew the third part of the stars of heaven, and did cast them to the earth: and the dragon stood before the woman who was ready to be delivered, for to devour her child as soon as it was born.

Currently Satan still has access to the glorified realm of the first heaven of the Earth (mentioned earlier when discussing the first heaven), and most likely every cursed world throughout the universe he managed to deceive during his rebellion, just as he had obtained dominion over Earth by deceiving Adam and Eve. His access will eventually be restricted to only Earth (Revelation 12:9; 12–13), at which time he will have his greatest level of authority over Earth he will ever achieve, through the person of the Antichrist. Fortunately this will only be a short period (about seven years) (Daniel 9:26–27).

Revelation 12:9–12

And the great dragon was cast out, that old serpent, called the Devil, and Satan, which deceives the whole world: he was cast out into the earth, and his angels were cast out with him. And I heard a loud voice saying in heaven, Now is come salvation, and strength, and the kingdom of our God, and the power of his Christ: for the accuser of our brethren is cast down, which accused them before our God day and night. And they overcame him by the blood of the Lamb and by the word of their testimony; and they loved not their lives unto the death. Therefore rejoice ye heavens, and ye that dwell in them. Woe to those who inhabit the earth and of the sea! For the devil is come down unto you, having great wrath, because he knows that he hath but a short time.

Daniel 9:26–27* (bracketed comments added)

And he shall confirm the covenant with many for one week [seven years]: and in the midst of the week he shall cause the sacrifice and the oblation to cease, and for the overspreading of abominations he shall make it desolate, even until the consummation, and that determined shall be poured upon the desolate.

*Each day of Daniel's seventieth week represents one year.

Aside from fallen angels, scripture may provide insight into other beings lower than the angels (the humans of Earth) obtaining access to glorified realms as well. The culmination of events that led to the destruction of the Tower of Babel (Genesis 11) may be a documentary of humanity's earliest experiments with interplanetary/interdimensional travel. In my book *Aliens in the Bible*, I speculated that humanity might have developed a means of reaching into the heavens through a technique called astral projection. Astral projection is the act of separating one's spirit from one's physical body without dying.[26] If humans (cursed beings in this example) did actually obtain access to glorified realms (the heavens) as scripture indicates in Genesis 11 (with a literal interpretation), it clearly explains why God destroyed the knowledge of this ancient civilization. Any cursed being obtaining access to glorified realms can only invite corruption to said realms.

Genesis 11:4

And they said, Go to, let us build us a city and a tower, whose top may reach unto heaven; and let us make us a name, lest we be scattered abroad upon the face of the whole earth.

A tall building is no big deal in my opinion—why would God care about a tall building? The buildings of today by far exceed the estimated three-hundred-foot-tall Tower of Babel.[27] I think something else was at the root of God's actions in this passage of scripture. The Tower of Babel was a Ziggurat (terraced pyramid temple), and the means by which it reached into the heavens was through a spiritual gateway.

Anthropologist Carlos Castaneda speaks of performing astral projection, under the apprenticeship of Iroquois sorcerer don Juan Matus. Astral projection is taught in numerous cults, and sometimes induced by the use of specific drugs, such as peyote or mescaline. Shamanism is one of the oldest religions in the world, dating back to the most ancient of civilizations, even to that of ancient Babylon,[28] which happens to be the same location as the Tower of Babel.[29] For this reason, I hypothesize that the means by which people reached into the heavens in Genesis 11 was not through the height of the Tower of Babel, but rather through the attainment and practice of spiritual knowledge that these people were not supposed to have.

In reference to the Tower of Babel, and Nimrod specifically (the king of that region), many Christian archeologists have concluded that they believe Nimrod (whose name wasn't translated as a proper name, but rather a reference to his character, because it means "rebel" in Hebrew), was actually the same hero mentioned in the Epic of Gilgamesh. In an article found on Christiananswers.net, Bible archeologist Dr. David P. Livingston explains a great deal about the Epic of Gilgamesh, which is an ancient, twisted version of the Biblical account of Creation, the flood of Noah, and some of the events that transpired shortly thereafter.[30]

While the Epic of Gilgamesh supports polytheism and the acceptance of sexual immorality as a norm, among other things, it may provide a historical glimpse into the activities occurring around the Tower of Babel, which stirred God to complete indignation. To begin with, Gilgamesh (Nimrod) may have been a Nephilim (part human, part fallen angel, as the Bible mentions in Genesis 6). Following is an excerpt from Dr. Livingston's article on Christiananswers.net:

> The Epic of Gilgamesh has some very indecent sections. Alexander Heidel, the first translator of the epic, had the decency to translate the vilest parts into Latin. Spieser, however, gave it to us straight.[31] With this kind of literature in the palace, who needs pornography? Gilgamesh was a vile, filthy, man. Yet the myth says of him that he was two-thirds god and one-third man.[32]

With this in mind, it should be clear to understand God's anger against the inhabitants surrounding the Tower of Babel. Putting these pieces of information together, one could postulate that they were astral projecting from the Tower of Babel and reaching into the heavens (or at least into the first heaven). Once there, they made contact with beings there, and reinitiated the events that led to the Nephilim in Genesis 6. God's statements in Genesis 11:6 concerning the inhabitants of the Tower of Babel appear to indicate that these people were the instigators in their own corruption, rather than fallen angels being the instigators. One fact is certain in any case; the events of Genesis 6 leading up to the Nephilim continued even after the flood of Noah (Genesis 6:4), and the Tower of Babel was most likely one of the earliest epicenters of this activity. One way or another, established boundaries were being crossed which violated a heavenly ordinance.

> Genesis 11:6 (bold emphasis added)
>
> And the LORD said, Behold, the people is one, and they have all one language; and this they begin to do: and **now nothing will be restrained from them**, which they have imagined to do.
>
> Genesis 6:4* (bold emphasis and bracketed comments added)
>
> There were giants in the earth in those days; **and also after that**, when the sons of God [fallen angels] came in unto the daughters of men, and they bare children to them, the same became **mighty men which were of old, men of renown**.
>
> *The phrase, "those days," refers to the days of Noah, and "after that," refers to the days that followed Noah, which must have followed the flood of Noah as well.*

Finally my last example of cursed beings obtaining access to glorified realms takes a bit of an imagination to contemplate, but if everything I have written thus far is taken into consideration, the prospect is conceivable. As touched upon earlier, the disincarnate spirits of Earthbound ghosts, demons, and devils were listed as examples of beings who have somehow crossed over an interdimensional barrier to be perceived by physical beings—and in some cases, even possess them. While most people might limit their thinking of disincarnate spirits to Earth only, one might consider the possibility that perhaps disincarnate beings may be capable of not only crossing interdimensional barriers, but interplanetary barriers as well.

In my book *Aliens in the Bible*, I discuss the possibility of Satan's domain encompassing Earth and other planets in our current galaxy, to include a possible angelic kingdom on Mars, and the planet Astera (some astronomers postulate Astera was a planet that once existed, but exploded and became the asteroid belt). If ancient civilizations (composed of angels and beings who reproduce but who have not been translated into angels yet) once existed in these places, then became corrupted under Satan's rule and were destroyed, the spirits of the nonangelic beings might have gravitated to Earth (or were taken to Earth by force). I say this because Earth contains within it the prison originally intended for Satan and his band of fallen angels—Tartarus. Then again maybe the angelic prisons within Earth are for those who sinned on Earth. It could even be that every planet containing beings who were deceived has inner prisons similar to the inner realms of Earth, and when Earth is eventually purged, and death and hell are thrown into the lake of fire, all of these prisons scattered throughout the universe will be thrown into the lake of fire as well. This would make sense because all the lower realms of hell except for the lake of fire are temporary. Could this be part of what the Bible means when it says there will be a "new heavens" and "new Earth?" (Isaiah 65:17; Isaiah 66:22; 2 Peter 3:13). Who knows?

Isaiah 65:17

For, behold, I create new heavens and a new earth: and the former shall not be remembered, nor come into mind.

So can people be possessed by the disincarnate spirits of other species who came from other planets? Such a prospect might actually be possible.

5.2. Interplanetary/Interdimensional Travel Conducted by Glorified Beings

Glorified beings who are faithful to God not only have access to other worlds—they require it in order to meet God's mandate that they should serve God throughout the heavens (Hebrews 1:14). How could the war in the heavens ever have been conducted, were it limited to only a single planet? How could angels help humanity if they couldn't even reach Earth from their home worlds?

Hebrews 1:14

Are they not all ministering spirits, sent forth to minister for them who shall be heirs of salvation?

Among the glorified beings faithful to God who access other worlds, some may be so powerful they simply will themselves to other planets at the speed of thought. But I would consider this type of power reserved for a very select few angels, if any at all. The Bible is full of examples of angels appearing to use technology (chariots of fire, clouds of glory, the Star of Bethlehem, and so on), which is why I am of the opinion that most angels use technology to conduct interplanetary/interdimensional travel.

As for God, he can certainly will himself to any location in time and space without the assistance of anything whatsoever (Acts 17:24–25), but interestingly, he frequently chooses to come to Earth in the company of some of his angels, with evidence of technology as well. Ezekiel, chapters 1 and 10, and numerous other scriptures appear to be accounts of such visitations.

> Acts 17:24–25 (bold emphasis added)
>
> God that made the world and all things therein, seeing that he is Lord of heaven and earth, dwelleth not in temples made with hands; Neither is worshipped with men's hands, **as though he needed any thing**, seeing he giveth to all life, and breath, and all things…

About two years ago, I used to correspond with author Frank E. Carlisle, who wrote *Solving the Riddle of Ezekiel's Wheels: The Chariots of the Cherubim/The Chariots of God*.[33] His exposition of Ezekiel's encounter is very in depth; my only critique on his work had to do with how he handled his conclusion. He did well in proving that Ezekiel's encounter involved a spacecraft, but after proving this, he mentions nothing about the implications of the findings. He even went on a tour of lectures at various universities throughout the United States, proving that Ezekiel encountered aliens, but then left the bewildered young students swimming with questions. Addressing the implications of the existence of extraterrestrial life is more important than proving they exist, in my opinion. Once people understand that alien spacecraft may be associated with God, they must also understand that God is still God, and not some alien being! Jesus himself ascended to heaven and disappeared into a cloud when he left the Earth (Acts 1:9), most likely to accompany angels in a spacecraft, but that doesn't mean he's merely an alien from another world.

> Acts 1:9
>
> And when He had spoken these things, while they beheld, He was taken up; and a cloud received him out of their sight.

I personally think it's somewhat funny because I know that God doesn't need any chariots of fire or special clouds. They're obviously his chosen mode of transportation in some instances, for reasons other than his dependence on them. Suggesting his dependence on anything, other than himself, is nonscriptural. He created the universe by speaking, so why would he need a gadget to go from place to place? God is God, and he *needs* absolutely nothing because his power is limitless. His angels, on the other hand, have their limits. Most likely God simply accompanies his angels, who depend on technology in some instances, just as he limited himself to human flesh when he walked the Earth, and rode on the back of a donkey (Matthew 21:7). Did he need that donkey? No, he could have floated into town surrounded by a procession of angels if he wanted to, but he didn't. In effect, the vehicle's association with God is irrelevant when it comes to God's identity and attributes.

Unfortunately many UFO researchers have come to illogical and deceptive conclusions regarding God's affiliation with what the Bible describes as an assortment of various vehicles that traverse the skies (clouds of glory, chariots of fire). UFO researchers who use the Bible as a reference to ancient accounts of alien contact may be accurate about *some* of the conclusions they have arrived at, but they couldn't be more wrong to demote God to the status of simply another created being, as many of them have. As stated earlier, I've read a number of books and heard endless drivel from people who claim that God is nothing more than a powerful alien—and that the human race is nothing more than a genetic experiment. Zechariah Sitchen, for example, the author I mentioned near the beginning of this book, supports the view that the plural word for God in Genesis 1:26, being Elohim, refers not to the Father, Son, and Holy Ghost, but rather a *race* of advanced alien beings. Like many UFO/alien book authors, Sitchen goes far in proving the existence of extraterrestrial life, but his grand conclusion leads to deception.

> Genesis 1:26 (bracketed comments added)
>
> And God [Elohim, plural form for God, referring to his Trinitarian nature] said, Let us make man in our image, after our likeness: and let them have dominion over the fish of the sea, and over the fowl of the air, and over the cattle, and over all the earth, and over every creeping thing that creepeth upon the earth.

Brad Steiger, another best-selling author, documents the experiences, and views of many alien abduction victims, yet remains silent concerning the implications of his reports. Consider the following excerpt from his book, *The Fellowship: Spiritual Contact Between Humans and Outer Space Beings*:

> As with many UFO contactees, Moi-ra and Ra-Ja Dove are convinced that the Aquarian Age is heralding in a new religion. Even the word religion will not be used anymore. For the main crux of the matter will have a much deeper sense of reality. The person will evolve out of believing in something into becoming something. The person will know what the religions of the ancient age have

always tried to demonstrate: to be still and know that you are god! Indeed, this is the great new religion! Each and every person will know that he is god.³⁴

Where have I heard this quote before? Oh yea, Psalm 46:10 quotes God as saying "Be still, and know that I am God..." I think God wants us to know that *he* is God, not *us*. He mentions the phrase "know that I am the Lord" seventy-seven times in the Old Testament. Is that not enough? Genesis 3:5 quotes the **devil** as contradicting God in this matter, and Satan's quote lines up precisely with the deception I quoted above from Brad Steiger's book. I count these deceptions as a precursor to the delusion spoken of in 2 Thessalonians 2:11.

Psalm 46:10—Quote from God (bold emphasis added)

Be still, and **know that I am God**: I will be exalted among the heathen, I will be exalted in the earth.

Genesis 3:6—Quote from Satan (bold emphasis added)

For God doth know that in the day ye eat thereof, then your eyes shall be opened, and **ye shall be as gods**, knowing good and evil.

2 Thessalonians 2:11–12

And for this cause God shall send them strong delusion, that they should believe a lie: That they all might be damned who believed not the truth, but had pleasure in unrighteousness.

Going back to the *Star Trek* theme I used to introduce my concept of the heavens, there is a "prime directive" of sorts regarding interspecies conduct. For those familiar with Genesis 6, I consider it an obvious conclusion that the sons of God mentioned there, which were actually fallen angels, violated this prime directive. (For those unfamiliar with *Star Trek* and the prime directive, the prime directive is a policy of noninterference mandated by intergalactic explorers belonging to a Federation of Planets. The policy of noninterference was an issue of ethics based on the anthropological premise that interference of technologically advanced species into the lives of less technologically advanced species may create more problems than it would solve.)³⁵ Clearly interspecies interaction includes something not mentioned in *Star Trek*—the will of God. In fact, the only true prime directive that exists can be reduced to one thing: God's will.

God mandates that the angels are to protect humanity (Hebrews 1:14) and learn of God's grace through their observations of humanity (I Peter 1:12; Ephesians 3:10). To this extent, they may even penetrate human society in disguise (Hebrews 13:2).

Hebrews 1:14

Are they not all ministering spirits, sent forth to minister for them who shall be heirs of salvation?

1 Peter 1:12

Unto whom it was revealed, that not unto themselves, but unto us they did minister the things, which are now reported unto you by them that have preached the gospel unto you with the Holy Ghost sent down from heaven; which things the angels desire to look into.

Ephesians 3:8–10 (bold emphasis added)

Unto me, who am less than the least of all saints, is this grace given, that I should preach among the Gentiles the unsearchable riches of Christ; And to make all men see what is the fellowship of the mystery, which from the beginning of the world hath been hid in God, who created all things by Jesus Christ: To the intent that **now unto the principalities and powers in heavenly places might be known by the church the manifold wisdom of God**...

Hebrews 13:2

Be not forgetful to entertain strangers: for thereby some have entertained angels unawares.

They are furthermore authorized to intervene in the affairs of humanity (which differs greatly with *Star Trek*'s anthropological emphasis of no intervention whatsoever), for which there is an abundance of scriptural support. Just to mention a few of the angels' interventions: they determine the outcomes of wars (Assyrian army destroyed, among many others—2 Kings 19:15), dish out God's wrath (destroyed Sodom and Gomorrah, sent plagues against Egypt, and will send many more throughout the entire world in the end times—Genesis 19:28; entire book of Exodus; Revelation 7–20), and serve as divine messengers (visions of the future—Daniel 8:16; 9:21; announcement of the birth of Christ—Luke 1:19, 26, and many others). For those interested in angelic intervention of a smaller, more personal nature, I recommend *A book of Angels*, by Sophy Burnham. In chapter 2, she mentions a few interesting true stories about people encountering angels. As she puts it, "When you need something, something makes it appear."[36] This was the case for a boy who cut his finger to the bone and was all alone at home, when suddenly a nurse knocked on the door, patched the boy up, then left without leaving any clues as to her identity or purpose for her visit. Then there was the car with a bald tire, which prevented the same boy and his mother from visiting his grandmother in Massachusetts; they didn't have enough money to buy a replacement. While the boy was explaining to a friend of his that they didn't have enough money to buy a new tire, a lady on the top of a hill nearby his house appeared from out of nowhere, and rolled a tire down the hill, dumping it into a shallow stream next to his house. Most people have interesting stories like these, and they frequently include a mysterious stranger who helped in a time of need. These stories are especially prevalent with children. I personally can testify to this, having seen the face of death a number of times, and each time, I was spared.

- I was sucked into a whirlpool during a camping trip—three people formed a chain to pull me out.
- I almost went over a waterfall during another camping trip—my brother saved my life.
- I was buried alive with my brother when we were digging a tunnel in the side of the riverbank near my house. Fortunately, my left arm was close enough to the surface for me to dig my own head out and breath! I then dug my brother out part way, but he was pinned by a massive boulder. I flagged a car down at a nearby road, and four muscular men pulled the boulder off my brother. (The same boulder rolled right over the top of me before landing on my brother, but the sand somehow cushioned me from its crushing weight.)
- I nearly fell off a cliff while climbing down a rope to get inside a cave. I don't know how I made it, but I did.
- I was nearly bit by a water moccasin in the pond by my house in Arkansas.
- I almost stepped on a rattlesnake during a camping trip in California, and almost stepped on a copperhead during another excursion in the woods in Arkansas. Both snakes were extremely large.
- I survived double pneumonia and bacterial meningitis. (Bacterial meningitis is the bad one that can cripple a person for life, but I recovered unscathed.)

Strangely, these brushes with death didn't end with my childhood, though that's where I encountered most of them. I was a few inches from driving off a bridge and landing in a river a few years ago, and just last year, I was a few inches away from becoming someone's hood ornament while riding my bike home from work. Someone ran a red light, and missed me by less than an inch; I physically touched the car as it passed me.

After reviewing this list, which is not all-inclusive, I feel as if someone wants me dead, but someone else who is more powerful wants me alive! I'd be a fool if I didn't think angels were helping me.

Angels intervene for the inhabitants of Earth as a whole, for nations, cities, families, and individuals. They almost never identify themselves, preferring to defer the glory to God, and they're usually gone before anyone realizes something out of the ordinary just occurred. Clearly, they don't solve all of our problems the way we might sometimes think they should, because the Earth is rife with turmoil. However, in all cases of interspecies interaction, angels who are faithful to God are conducting their actions in strict accord with the will of God.

Part II
The Heavens of the Past

Scripture states in Genesis 1:1, "In the beginning, God created the heavens and the Earth." Scientists speculate that this occurred anywhere from 4.5 to 17.5 billion years ago, depending on a number of variables.[1] When I wrote *Aliens in the Bible*, I had various contentions with the old-Earth theory. I hadn't fully studied all of the scriptural support behind it because I thought that proponents of this theory were trying to make the theory of evolution jive with the Bible. Some do, but not all.

I also couldn't conceive of a logical explanation for creatures like dinosaurs, which could only exist in a world that had sin and death—to my knowledge, it was Adam and Eve who brought sin and death into the world. I have since discovered that there is an explanation as to how sin and death could have entered the world prior to the creation of Adam and Eve. Now that I am familiar with the scriptural arguments of the old-Earth theory (which includes the existence of a pre-Adamite, partially angelic civilization, which I will discuss shortly), I have actually flip-flopped on my interpretation of the creation account in Genesis. Some may think I am hurting my own credibility by admitting that I have changed my views on previously researched material, but these same people may also take note that I am in accordance with my own admission. As I have said in the past, I proclaim once again: in anything I write, if someone can produce an effective scriptural rebuttal to persuade me otherwise, I am willing to recant my position.

Continuing with the dialogue of creation, after creating the heavens and the Earth, God filled the heavens with life (the host of heaven). This was the age known as the dispensation of angels; it began a long time ago, before humanity existed. For a very long time, all the host of heaven were obedient to God, and the universe dwelled in perfect harmony (Job 38:4–7).

Job 38:4–7

> Where were you when I laid the foundations of the earth? Declare, if thou hast understanding. Who hath laid the measures thereof, if thou know? Or who hath stretched the line upon it? Whereupon are the foundations thereof fastened? Or who laid the corner stone thereof; when the morning stars sang together, and all the sons of God shouted for joy?

As a side note, when God created life in the universe, I suspect he created all the species to reproduce after their own kind, just as he did with all the species he created on Earth. Some people might disagree with me on this point, but I think scripture provides evidence to support this possibility. First of all, all the species of Earth bear this reproductive trademark, including humans, some of which will eventually be translated into angels (Matthew 22:30; Mark 12:25; Luke 20:35; and Revelation 21:16, which I will explain more about later). Why would angels who were formerly human be any different in this respect from angels who come from another species?

Second, the account of Genesis 6 documents the fact that angels are capable of reproduction. I'm not alone in this interpretation. Finis Jennings Dake, author of *Dake's Annotated Reference Bible, Revelation Expounded, Bible Truths Unmasked*, and *God's Plan for Man*, pastor of several churches, founder of a Bible school, lecturer, and radio broadcast host for thirteen years, understood with exceptional clarity that the sons of God documented in Genesis 6 were of unearthly origin. Other authors, such as Chuck Missler (former CEO of Western Digital Corporation, Branch Chief of the Department of Guided Missiles, and currently the president of Koinonia House Ministries) and Dr. Mark Eastman, MD, both Christians devoted to ministry, understand who the Nephilim are as well. I highly recommend Chuck's tape series, *Return of the Nephilim*, and the book that Chuck and Dr. Eastman wrote together, *Alien Encounters: The Secret Behind the UFO Phenomenon*.

While it may be possible for angels to assume a reproductive form, I consider it an unlikely scenario for a being who would be foreign to such activities. The fallen angels of Genesis 6 were beings committing acts of sensual lusts, which I think would be extremely odd, had they not been reverting to a former nature with which they were familiar. There would simply be no motive to partake of such deeds. I only digress in this matter because knowing and understanding this aspect of angels may help people to understand their past and possible future interactions with humanity.

I've heard it said that angels are only male, but I consider this argument weak because it is an argument from silence: angels are never referred to in scripture as being female, so the absence of said references makes the argument. But their existence is not disproved simply because they aren't mentioned in the female gender. Jesus indicated that humans will one day be angels (Matthew 22:30; Mark 12:25; Luke 20:35), and many humans are female! These facts put together means that there will be female angels.

Matthew 22:30

For in the resurrection they neither marry, nor are given in marriage, but are as the angels of God in heaven.

Mark 12:25

For when they shall rise from the dead, they neither marry, nor are given in marriage; but are as the angels which are in heaven.

Luke 20:35

But they which shall be accounted worthy to obtain that world, and the resurrection from the dead, neither marry, nor are given in marriage: Neither can they die any more: for they are equal unto the angels; and are the children of God, being the children of the resurrection.

Scripture was also influenced by the culture of the times in which it was written; literature of biblical times always defaulted to the male gender. Most languages in the world are still this way today. According to the Wikipedia Encyclopedia, "In Indo-European and Afro-Asiatic languages, male pronouns have traditionally been used when referring to both genders or to a person or people of an unknown gender."[2] In the male-dominated society of biblical times, it would have been natural for the authors of the Bible to refer to the beings who were "greater than men" as being male rather than female (Psalms 8:5; Hebrews 2:7–9). Galatians 3:26–28 is a prime example of the influence of culture in literature; this scripture calls all Christians "sons of God through faith in Christ Jesus," in the English Standard Version of the Bible, and this includes women because it states immediately afterward that there is "neither male nor female, for you are all one in Christ Jesus." Such an admission was extremely liberal for the day in which it was written, yet even with this bold statement of equality between men and women, all are referred to as sons. Scripture is scattered with numerous examples like this that document the simple fact that the literature of biblical times (and even today), frequently defaults to the male gender when speaking in general terminology.

So God created the universe and filled it with life. Then after some time had passed, various species advanced greatly (advanced, not evolved). These species completely populated their planets, then spread to other planets. God translated many of these advanced species into new natures that no longer reproduced, just as he will do for humanity in the future (Matthew 22:30; Mark 12:25; Luke 20:35; Revelation 21:16, 22:8–9). This new nature differentiates angels from the larger group I've been referring to as the host of heaven. Mortal humans, for example, are considered hosts in scripture (Exodus 12:41), whereas glorified, translated humans are called angels (Revelation 22:8–9).

THE HEAVENS OF THE PAST   59

Exodus 12:41

And it came to pass at the end of the four hundred and thirty years, even the selfsame day it came to pass, that all the hosts of the LORD went out from the land of Egypt.

Revelation 22:8–9 (bold emphasis added)

And I John saw these things, and heard them. And when I had heard and seen, I fell down to worship before the feet of the angel which showed me these things. Then said he unto me, See thou do it not: for **I am thy fellow servant**, and **of thy brethren the prophets** and of them which keep the sayings of this book: worship God.

There are various classes of angels throughout the universe (Ephesians 6:12; Colossians 1:15–16); the most powerful among them are called archangels (Daniel 12:1; Revelation 12:7–9—Michael is the archangel who defeated Lucifer and cast him out of the highest heaven, which means that Lucifer was probably an archangel as well). In the Old Testament, archangels are called *gods* in several places, but the ancient Israelites clearly understood the distinction between these less powerful, finite, created beings, and their Creator, the eternal God, who has unique attributes (all powerful, all knowing, no beginning or ending, the Creator, and so forth) that no created being will ever have.

Ephesians 6:12 (bold emphasis added)

For we wrestle not against flesh and blood, but against **principalities**, against **powers**, against the **rulers** of the darkness of this world, against **spiritual wickedness in high places**.

Colossians 1:15–16 (bold emphasis added)

Who is the image of the invisible God, the firstborn of every creature: For by him were all things created, that are in heaven, and that are in earth, visible and invisible, whether they be **thrones**, or **dominions**, or **principalities**, or **powers**: all things were created by him, and for him...

Daniel 12:1 (bold emphasis added)

And at that time shall **Michael** stand up, **the great prince which stands for the children of thy people**: and there shall be a time of trouble, such as never was since there was a nation even to that same time: and at that time thy people shall be delivered, every one that shall be found written in the book.

Revelation 12:7–9 (bold emphasis added)

And there was war in heaven: **Michael and his angels fought against the dragon**; and the dragon fought and his angels, and prevailed not; neither was their place found any more in heaven. And the great dragon was cast out, that old serpent, called the Devil, and Satan, which deceives the whole world: he was cast out into the earth, and his angels were cast out with him.

Lucifer once held the title of "son of the morning" or "Day Star" in the English Standard Version (Isaiah 14:12). He was given dominion over the Earth long before humanity ever existed (Isaiah 14:12–14; Ezekiel 28:11–17; Luke 10:18; 2 Peter 3:4–8; Jeremiah 4:23–26; as well as many other passages that I will discuss later in this book). He had access to many other worlds: in Ezekiel 28:14–16, "stones of fire" most likely represent solar systems in the second heaven (heavens), because Satan being restricted from them is linked to the overthrow of his kingdom. I will now discuss these concepts in detail.

3

The Earth Prior to the Creation of Adam and Eve

Discussing Earth's history predating Adam and Eve sheds light on a number of mysteries. Paramount among these mysteries is the modern enigma of alien lifeforms, an enigma that actually isn't modern at all. From the Bible research I've conducted on Earth's ancient past, I now understand the underlying reason why aliens (which I consider to be synonymous with the host of heaven, some of which are angels), have been visiting Earth for so long. I also know why those who are hideously evil and far more powerful than humanity haven't invaded Earth in force—at least not yet (because of the restraining power of the prayers and actions of the saints of the church—2 Thessalonians 3:6). Furthermore I have a new view on the vast scope of salvation, and I can see what role alien contact may play in Earth's future in regards to the scope of salvation. But before I discuss the grand conclusion of all these findings, I must first return to an ancient Earth, which predates humanity.

As I already mentioned, I once believed that Earth was much younger than the widely accepted scientific view currently suggests. I believed Earth was young for one main reason: the Bible provides detailed genealogical information dating all the way back to Adam that accumulates to roughly six thousand years. Because Adam was created on the sixth day of creation, the Earth is only a few days older than Adam is.

Contradicting this calculation is the fact that dinosaurs have been carbon dated to be anywhere from 66 to 245 millions of years old.[1] But how could dinosaurs exist in a world where there was no sin or death? It was through Adam and Eve that sin and death were introduced into the world, so dinosaurs, being the vicious killing machines that they were, couldn't have preceded Adam and Eve. No carnivores could exist prior to Adam and Eve—or so I thought. Because of these facts, I concluded that Earth must be much younger than the majority of the scientific community believes, but my views on this have since changed.

I recall how fascinated I was in early 1996, when I first stumbled across the Watcher Web site.[2] When I read about ancient angelic civilizations that once existed on Earth prior to Adam and Eve being created, it was as if a light turned on in my spirit. Something about it seemed right. That light quickly faded, however, when I systematically picked apart the researched material I was reading. Most of the scriptural references seemed to be taken out of context, concerning the existence of a pre-Adamite civilization on Earth.

When it comes to Bible research, I'm very systematic about how I interpret scripture. I always take the literal meaning first because the Bible is primarily a simple, no-nonsense document. If it's from God to his children, then the majority of it should be easy to understand at face value. Also the literal interpretation is always the easiest interpretation, and not as subject to opinions as symbolic interpretations, so it's the most stable interpretation.

On the other side of the coin, however, if scripture is from God, then it may very well contain a nearly infinite degree of meaning, hidden as nuggets of truth. As scripture states, it is the honor of kings to seek them out (Proverbs 25:2).

Proverbs 25:2

It is the glory of God to conceal a thing: but the honor of kings is to search out a matter.

In short, scripture provides the light we need, and it dynamically fills our spirits like water in a sponge, uniquely tailored to each person's level of maturity, intellectual astuteness, spiritual wisdom, and a vast array of other characteristics that make us the individuals that we are.

In most cases, scripture provides a clearly understood literal meaning, but often embedded deeper within are numerous symbolic meanings. Most scriptural interpretative mistakes are made when a possible symbolic meaning is given more weight than the literal meaning. This is a common theme among most cults, sometimes to the degree that a symbolic meaning might even contradict the literal meaning. This only happens, of course, when the human imagination is playing on the interpretive Ferris wheel—without the company of logic or common sense, and especially without the Holy Spirit's illumination. Symbolic interpretations must always be consistent with the rest of scripture, and the context of the passage. These basic rules have helped me understand the Bible immensely, but I still had some learning to do back then, just as I do now.

The Earth Prior to the Creation of Adam and Eve 63

Just to give a simple example of how symbolic interpretations of scripture can distort the true meaning, the doctrine of reincarnation is directly refuted with a literal interpretation of Hebrews 9:27. Furthermore, the message of justice combined with grace, mercy, and salvation brought through the sacrifice of Jesus on the cross would have little meaning if unrepentant people that died in sin were just reincarnated. Jesus would have died for nothing if this was the case, yet there are still people who insist that Luke 1:17 promotes the doctrine of reincarnation, when it clearly doesn't.

The whole point of streaming off on this tangent about how I interpret scripture is simply to say that the Watcher Web site I previously alluded to wasn't written well enough for me to accept the ideas that were espoused. The site spoke of how the account of creation in Genesis starting at Genesis 1:2 is actually the re-creation of Earth to a habitable state—and furthermore that an ancient angelic civilization existed on Earth prior to the creation of Adam and Eve. I was initially fascinated, but when I looked up the scriptural references on the site, they appeared to be taken out of context.

When I wrote *Aliens in the Bible*, I used information from the Watcher site to provide supporting information concerning a possible ancient civilization on Mars. I spoke of Richard Hoagland's exploded planet hypothesis, which suggests that there was once another planet in our current solar system (known as Astera) that exploded and became what is now the asteroid belt. The Bible refers to the destruction of this planet in several places, which I documented in *Aliens in the Bible*. For the sake of brevity, I will only repeat the scientific discoveries of that same research here, yet link them with new relevant Bible passages that I didn't mention in *Aliens in the Bible*.

Concerning Mars in particular, there are four items of scientific interest that have developed in recent years indicating evidence of life that once may have existed on Mars. These include the discovery of organic compounds found on asteroids, what appears to be fossilized microorganisms found on a meteorite from Mars, the face and pyramids also found on Mars, and evidence of a cataclysm that destroyed the atmosphere of Mars.

About the first item of interest, the discovery of organic compounds found on asteroids, the 1987 October edition of Science magazine had this to say:

> D.P. Cruikshank and R.H. Brown reported a startling piece of news. They had discovered organic compounds on three asteroids: Murray, 103 Electra and Orgueil. Utilizing the process of spectral analyses of reflected light from these three asteroids, Cruikshank and Brown detected amino acids. More startlingly, "aqueous alteration products," such as clay were found, suggesting that

the parent body had been affected by water. If these asteroids did in fact contain sediment, it could not have deposited without large quantities of water. But these were asteroids—relatively minute chunks of rock hurtling around the sun from a common area between the orbit of Mars and Jupiter. The evidence found on these asteroids could only mean that they were from a parenting body possessing an atmosphere and oceans.[3]

As for the meteorite from Mars, NASA's report regarding possible fossilized evidence of life is reported on a government Web site as follows:

> In August 1996, a group of scientists announced that they had found evidence of ancient life on Mars. This evidence included bacteria-shaped objects and organic chemical molecules in the Martian meteorite ALH 84001, which was collected in Antarctica. In the next few days, NASA presented the work at a press conference, the President made a statement about it, and the TV and papers were full of reports, speculation, and jokes about life on Mars.[4]

Coupled with the first two discoveries above, photographs of the face and the pyramids on Mars have been discussed in depth in the media. Widely disputed as tricks of light and shadows, the photos of what appear to be artificial structures on Mars are still compelling.[5]

Further documenting the history of Mars, the above article continues.

> Satellites sent to Mars in 1976 collected information concerning the geological nature of Mars, and its atmosphere. The images from the orbiters mapping sequence made it clear that Mars had experienced a nearly unimaginable catastrophic episode. With the evidence of oceans of water having once flowed on Mars' surface in huge quantities, it was apparent that the Martian atmosphere was once more dense, and the climate much more hospitable. Sometime in the remote past, for reasons still being debated by astrophysicists, there was a cataclysm on Mars. The Martian oceans washed over the surface of the planet, inundating their continents. The vast atmosphere was ripped away, and the once Earth-like environment was laid waste.[6]

What do organic compounds found on asteroids, microorganisms found on Martian meteorites, the pyramids and face on Mars, and an apparent catastrophe that occurred on Mars, all point to? All these clues can be wrapped up in one package with one unique theory, called the exploded planet hypothesis.

According to noted astronomer Tom Van Flandern in his presentation *Exploding Planets & Non-Exploding Universes: The mechanism for explosion*, the exploded planet hypothesis was very popular back in the 1950s. A scientist by the name of Ramsey came up with a number of ways in which terrestrial-sized planets could either implode or explode with changes of state with certain elements in the core. For example, one such change of state would be the turning of water into ice. According to Ramsey, if the pressure and temperature conditions were right, they might produce a change in state in a planet's core, which could result in a spontaneous explosion or implosion.[7]

Evidence of such a planetary explosion, according to conspiracy theorist and author Richard Hoagland, is exactly what comprises the asteroid belt of our solar system today. Asteroids found to contain organic compounds, along with the unexplained catastrophe enshrouding the fate of what Mars once was, both testify to a planetary explosion that occurred long ago.

Following is an excerpt concerning the generally accepted view of how the asteroid belt was formed:

> The accepted theory for the creation of the asteroid belt is usually the failed planet accretion theory. This theory states that during the primordial beginning of the solar system, a planet which astronomers call Astera was forming in the place now occupied by the asteroid belt. Jupiter's gravitational influence on the incipient planet was too strong for it to solidify. Because of Astera's insufficient mass early in its development, it fragmented.[8]

What Richard Hoagland points out, concerning this accepted theory for the creation of the asteroid belt, is that according to new evidence, it simply can't be correct. As he puts it, "The facts are becoming more obvious…both the planet Mars and this mysterious parent body of asteroids once sustained oceans and atmospheres."[9]

In short, the most likely scenario explaining the asteroid belt and the fate of Mars is that at one time, there was a planet that formed (or more correctly, was created). This planetary body, Astera, was an intact planet containing water and life, rather than a cluster of debris that failed to accumulate into a planet. Furthermore, for some mysterious reason, this planet exploded, and became what is now known as the asteroid belt.

A clue in scripture pointing to this catastrophic event may be found in the Hebrew word "Rahab," (not to be confused with the prostitute mentioned in the Old Testament). This word means "boaster," and most interpretations of its destruction are thought to be figurative of God destroying the proud, which I agree with, but I think there's more to it. While Psalm 89:10 simply mentions God destroying Rahab, Psalm 87:4 speaks of Rahab as if it's a place, grouping it with the destruction of Babylon. Isaiah 51:9 goes further to link the destruction of Rahab with the destruction of Lucifer's ancient kingdom of the past, in the "ancient days, in the generations of old." Rahab, therefore, may actually be an ancient Hebrew word that obtained its definition, "boastful," from the character of the citizens of an ancient kingdom that was destroyed long before humanity ever existed.

> Psalm 89:10
>
> Thou hast broken Rahab in pieces, as one that is slain; thou hast scattered thine enemies with thy strong arm.
>
> Psalm 87:4* (bold emphasis added)
>
> I will make mention of **Rahab and Babylon** to them that know me: behold Philistia, and Tyre, with Ethiopia; this man was born there.
>
> *Many places in the Hebrew language became words with meanings that evoke memories of the places. Sodom, for example, means, "burnt" in Hebrew. Babylon also derives its name from the Hebrew word "Babel," dating back to Tower of Babel, and the "confusion" that God cast over its inhabitants. In order to maintain consistency in the above passage, if Rahab is interpreted as "boastful," Babylon should be interpreted as "confusion," but a simple reading of the text precludes this. Clearly, Rahab and Babylon are both physical places that God destroyed.*
>
> Isaiah 51:9* (bold emphasis added)
>
> Awake, awake, put on strength, O arm of the LORD; awake, **as in the ancient days, in the generations of old**. Art thou not it that hath **cut Rahab, and wounded the dragon**?
>
> *Take note that the destruction of Rehab is what causes the wounding of the dragon.*

During the explosion of Astera (which may very well be Rahab), Mars was buffeted, as well as Earth. This event threw the entire solar system into complete disarray.

When I wrote *Aliens in the Bible,* I agreed with the Watcher Web site up until this point. The Watcher site then stated that some fragments from this explosion also destroyed the ancient civilization that once existed on Earth prior to Adam and Eve. The impact of this meteor was devastating, kicking up so much debris that the entire Earth became enshrouded in darkness. While I found no refutation for the theories on the Watcher site about ancient civilizations found on

Mars, I refuted the concept of a pre-Adamite civilization on Earth because, as I mentioned earlier, the research appeared to be taken out of context.

Flashing forward a few years later, after I published *Aliens in the Bible*, I read a book called *God's Plan for Man*, written back in the fifties by well-known theologian Finis Dake. *God's Plan for Man* was purchased as a gift for me by a complete stranger. This person read *Aliens in the Bible*, which is posted on my Web site, then purchased *God's Plan for Man* for me—a $50 book—and mailed it to my house. Thank you, Linda! Her investment in me has led to the extension of my research, which I wasn't planning at all.

I am now recanting my prior position refuting a pre-Adamite civilization on Earth. Finis Dake returned me to the ancient pre-Adamite history of Earth and proved with a very concise, systematic, contextual interpretation of the Bible that the existence of a pre-Adamite civilization is a credible possibility.

In *Aliens in the Bible*, I saw some of the pieces of the puzzle of Earth's ancient history when I discussed my findings of Satan's strange connection to Earth, which I never could quite figure out. Why, for example, would God cast Satan down to Earth, of all places, during the end times? This always puzzled me. If there are other intelligent species in the universe who were also deceived by Satan (Revelation 12:4), then what makes Earth any different from any other planet? The universe is a big place to single out one puny planet among countless others, for the ultimate showdown between good and evil. For any Christian to accept the possibility of the existence of extraterrestrial life, the question about why the Bible seems to be focused primarily on Earth, of all places in existence, must be answered.

I now have that answer.

Most Christians don't have a hard time with the perspective that Earth is the center of God's attention in the universe because most Christians think that Earth is the only planet in the universe with life on it. According to Tariq Malik, staff writer for Space.com, a telephone poll, which questioned one thousand Americans, revealed that 60 percent of Americans believe in extraterrestrial life. Interestingly, "Democrats and Republicans were equally likely to believe in life on other planets, while regular churchgoers were less likely to believe in extraterrestrial life (about 46 percent) than non-churchgoers (about 70 percent)..."[10]

Additionally, in May of 2005, a survey conducted by the National Institute for Discovery Science was given to pastors, priests, and rabbis across the United States, asking basic questions relating to extraterrestrial life.[11] The survey yielded some interesting results. Over 70 percent of the respondents felt that the existence of extraterrestrial life posed no threat to their fundamental beliefs, yet out of all of the comments, not a single one gave any indication of biblical knowledge concerning extraterrestrial life, as I espouse in this book. Not one of them connected the dots, forming the basic conclusion that angels, by definition, are intelligent beings whose origins are not from Earth. All of these people have been reading the Bible most of their lives, and couldn't see the basic fact that one of the primary themes of the Bible is God's intervention in the affairs of humanity through extraterrestrial beings.

Furthermore, of the 42 percent that gave positive responses, most of them had theological questions they thought might be a struggle to deal with, such as the concept of "original sin" existing in places other than Earth. Many of them also pondered the concept that humanity is favored above the other species in the universe (angels, among others) because we are made in the image of God. I, however, have yet to find a single passage of scripture that says that God actually *favors* humanity over angels, or other beings referred to as the host of heaven. I think the big picture includes humanity as another intelligent species, given a defined place of dominion (Acts 17:24–26), rather than the single most important species.

Acts 17:24–26 (bold emphasis added)

> God that made the world and all things therein, seeing that he is Lord of heaven and earth, dwelleth not in temples made with hands; Neither is worshipped with men's hands, as though he needed any thing, seeing he giveth to all life, and breath, and all things; And hath made of one blood all nations of men for to dwell on all the face of the earth, and **hath determined the times before appointed, and the bounds of their habitation**…

Furthermore, while it is true that humans are made in the image of God, angels are made in his image as well. Being made in God's image refers not to a physical or even a spiritual attribute, but rather to the function of ruling over a particular domain as God's representative. This is what the Hebrew word for image, "tsehlem," in context with the rest of the passage denotes (Genesis 1:26).[12] Humans were originally given dominion over the Earth, and other intelligent species, many of whom are angels, have dominion over their respective portions of the universe as well. That's why scripture states that humans will one day be equal with (but not greater than) the angels (Luke 20:34–36). In cases where angels are in the human domain, humans will have jurisdiction (1 Corinthians 6:3), but humans will most likely be under angelic jurisdiction when traveling through their domains.

Genesis 1:26* (bold emphasis added)

And God said, Let us make man in our image, **after our likeness: and let them have dominion** over the fish of the sea, and over the fowl of the air, and over the cattle, and over all the earth, and over every creeping thing that creeps upon the earth.

Notice that being made in the image of God, after his likeness, is directly linked to having dominion over Earth. Much could be speculated about what being made in God's image means, but the words that are clearly stated have to do with having dominion over Earth. In short, humanity was supposed to rule over the Earth as a representative of God, as God rules over the universe.

Luke 20:34–36 (bold emphasis added)

And Jesus answering said unto them, The children of this world marry, and are given in marriage: But they which shall be accounted worthy to obtain that world, and the resurrection from the dead, neither marry, nor are given in marriage: Neither can they die any more: for **they are equal unto the angels**; and are the children of God, being the children of the resurrection.

1 Corinthians 6:3

Know ye not that we shall judge angels? how much more things that pertain to this life?

Perhaps some angels rule entire solar systems, or even groups of solar systems, serving as God's representatives in those places in the heavens. Humans were ordained to rule the Earth, and one day when humanity extends beyond the confines of the Earth to populate other planets, their domain will increase to encompass those new worlds as well.

For those wanting to see further research into what it means to be created in God's image, I recommend reading *The Facade*, by author Michael S. Heiser.[13] While *The Facade* is a fictional UFO conspiracy story, Heiser is a student of several notable Semitic scholars, and he has an extensive linguistic background. *The Facade* is therefore an archive of true information relating to UFO phenomenon, as well as a miniature Bible commentary about UFO related information.

Heiser reviewed my book, *Aliens in the Bible*, and we corresponded in the past about various subjects, debating from time to time over matters of semantics mostly. While I agreed with him about his interpretation of what it means to be created in God's image, I did have a contention with him over another matter concerning the same passage of scripture listed above. He is of the opinion that God the Father, Elohim, is speaking to a *divine counsel*, the "us" in the passage, rather than God the Father speaking to other members of the Godhead (God the Son, and God the Holy Spirit). He covers this in depth in *The Facade*.[14] I think this is an important distinction, because God is the one doing the creating, not angels. To argue that angels were present and had any input in the matter of creating humanity blurs this distinction. Unlike Zechariah Sitchen, Heiser agrees that angels are not co-creators with God, but he still thinks they were present in their *divine counsel* when God was making decisions on how to create Adam and Eve. Heiser's reasoning behind this interpretation rests solely on the grounds of some linguistic rule that a concept covered in the New Testament, being the triune nature of God, can't be *read into* an interpretation of the Old Testament. My argument against this, however, is more of a commonsense approach. If God has a triune nature, then that nature was present during both the Old Testament and New Testament. I don't have to be a Semitic scholar to come to this conclusion; I just have to understand the Trinity. The Old Testament writers simply referred to God's triune nature by using the plural word for God, Elohim, because different writers use different words to describe the same things. (For anyone wanting to understand the Trinity more, I recommend reading about the Trinity in *God's Plan for Man*.)[15]

Another disagreement I have with Heiser concerns the matter of Lucifer's rebellion. He believes that Lucifer's initial rebellion against God began after Adam and Eve were created, when God gave humanity dominion over the Earth. There are several problems with this theory. First, scripture states that Lucifer had a throne before he sinned against God (Isaiah 14:12–14), which means that he was a king. This throne was also associated with the Garden of Eden. My question is, how could Lucifer be a king, if Adam and Eve had dominion over the Earth? The only way Lucifer could have a throne on Earth was if he was a king *before* Adam and Eve were created, which explains Lucifer's anger (because he lost his kingdom to Adam and Eve), but it also raises another question. What kind of a king rules an empty territory with no subjects? If Lucifer was a king, he had subjects, yet all this had to predate Adam and Eve, if Adam and Eve had dominion over the Earth.

Also, if one believes that Lucifer's *initial* sin was documented in Genesis 3:6, after Adam and Eve were created, then one must also believe that Lucifer managed to deceive one-third of all the angelic hosts of all the heavens (a countless number of beings according to Hebrews 12:22), in less time than it took Adam and Eve to have their first male child (Genesis 4:1). Furthermore, Lucifer not only deceived one-third of the inhabitants of the heavens, but he also rallied them together to go to war against God, and lead an attack into the third heaven. How could all this happen in just a few years at most? (If this information raises a bunch of unanswered questions, don't worry, because I will cover the topic of Lucifer's rebellion in depth later in this book.)

Heiser seemed to lack common sense in some of our discussions, giving more credence to professors rather than basic logic, at least in my opinion. He gave me the impression that he believed that I didn't truly understand the Bible because I wasn't a seminary graduate stuffed with theological trivia. Despite our differences, I still like Heiser, respect his work (and even reference it in this book), and look forward to meeting him someday, because we are both brothers in Christ.

Anyway, I apologize for digressing. Back to the topic, considering all of the above about angels ruling throughout the heavens, I have always wondered about the reason Earth is a focal point for God's activity in the universe, as scripture clearly depicts that it is. God picked Earth to initiate his great plan of attack against Satan and to save all the casualties of sin who want to be saved. To do this, God undertook a two-part mission, which is still in progress: come to Earth in the form of a human being (Jesus) and die for everyone's sins—then after conquering death, ensure the spreading of the Gospel of repentance and salvation.

Concerning the first part of his mission, I've heard it said that there couldn't be intelligent life on other planets (aliens) similar to humanity (beings dealing with sin and death) because Jesus would have to suffer the death penalty on each of those planets. My question is why? Scripture states that Jesus suffered death on the cross once and for all (Hebrews 10:10), and that's that. I see no reason he should have to suffer a similar fate anywhere else. Why would Jesus's dealing with sin and death be any different as another species on another planet? God defines sin and death with absolute definitions that pertain to all existence. Therefore the cross is sufficient for all of it, even if it extends into the cosmos.

Hebrews 10:10 (bold emphasis added)

By thee which will we are sanctified through the offering of the body of Jesus Christ **once for all**.

Concerning the second part of Jesus's mission, Hebrews 2:16–18 indicates that salvation is not for angels, but for the descendants of Abraham.

Hebrews 2:16–18 (bold emphasis added)

For verily He **took not on him the nature of angels; but He took on him the seed of Abraham**. Wherefore in all things it behooved Him to be made like unto His brethren, that He might be a merciful and faithful high priest in things pertaining to God, to **make reconciliation for the sins of the people**. For in that He Himself hath suffered being tempted, He is able to succor them that are tempted.

Does this confine salvation to the inhabitants of Earth? The true descendants of Abraham are not defined as biological descendants. This definition, based on one's DNA rather than the grace of God, would disqualify a large, faithful portion of even the human race; rather, the descendants of Abraham are those who belong to Christ (Galatians 3:26–29).

Galatians 3:26–29 (bold emphasis added)

For ye are all the children of God by faith in Christ Jesus. For as many of you as have been baptized into Christ have put on Christ. There is neither Jew nor Greek, there is neither bond nor free, there is neither male nor female: for ye are all one in Christ Jesus. **And if ye be Christ's, then are ye Abraham's seed, and heirs according to the promise.**

No word is mentioned about what species they are, or what planet or dimension they come from. Scripture only states that salvation is not for the angels, and even that statement is somewhat vague. What defines an angel? Better yet, *who* defines an angel?

Recapping what I've already mentioned, Jesus defined angels as beings who neither marry nor are given in marriage. He also said that humans who are saved would one day be translated into this new nature (Matthew 22:30; Mark 12:25; Luke 20:35; Revelation 21:16–17; 22:8–9).

Matthew 22:30 (bold emphasis added)

For in the resurrection they **neither marry, nor are given in marriage**, but are **as the angels of God in heaven**.

Mark 12:25 (bold emphasis added)

For when they shall rise from the dead, they **neither marry, nor are given in marriage**; but are **as the angels which are in heaven**.

Luke 20:35–36 (bold emphasis added)

But they which shall be accounted worthy to obtain that world, and the resurrection from the dead, **neither marry, nor are given in marriage: Neither can they die any more**: for they are **equal unto the angels**; and are the children of God, being the children of the resurrection.

Revelation 22:8–9 is a primary example of this glorified state, where scripture refers to a glorified saint (a man, one of the brethren) as being an angel (this is most clearly seen in the King James Version of the Bible). The apostles understood this to be the case as well; in Acts 12:12–16, when the apostles saw what they believed to be Peter's ghost (because they thought he had died in prison), they said it was Peter's *angel*.

<u>Acts 12:12–16</u> (bold emphasis and bracketed comments added)

> And when he had considered the thing, he came to the house of Mary the mother of John, whose surname was Mark; where many were gathered together praying. And as Peter knocked at the door of the gate, a damsel came to hearken, named Rhoda. And when she knew Peter's voice, she opened not the gate for gladness, but ran in, and told how Peter stood before the gate. And they said unto her, Thou art mad [everyone understood that Peter was currently in jail, so they thought]. But she constantly affirmed that it was even so. Then they said **It is his angel**. But Peter continued knocking: and when they had opened the door, and saw him, they were astonished.

Considering these details, as I already concluded before, one may easily deduce that the angels weren't always angels. They were most likely like us: created to reproduce after their own kind, just as Adam and Eve were until they reached a certain point in their existence when God predetermined to translate them into a new, higher level of existence. Also as previously mentioned, this theory explains a great deal about some of the behavior of angels, such as the fact that they are apparently physically attracted to humans (Genesis 6). Such attraction would be a very odd thing if reproduction were foreign to them.

Adam and Eve were never angels; they sexually reproduced. If Adam and Eve had never sinned, Earth would've eventually become fully populated. The next level of existence for humanity would have then commenced. At the proper time, God would have translated different people at different times into a new, nonreproductive nature, as we now know he still plans to do, despite the fall. I suspect that this is how God designed most, if not all, higher life-forms created in his image. Initially there is a reproductive stage, but then a translation takes place that mutates beings into an eternally glorified existence that no longer entails reproduction. In a sense, we are like butterflies—designed to go through a metamorphosis. This is where humanity has come from, where it is going, and—from what it looks like—where the angels, both faithful and fallen, have been. In essence to be an angel is to be translated by God into the highest form of existence, rather than to be made into an angel fresh from the start, as most Bible students think.

Putting all this together, one may surmise that salvation may not be for the angels because they are closer to God than any other being, and they have the highest level of accountability. There are probably countless numbers who aren't angels, however, who aren't from Earth or our current dimension. If they are in the same boat as the humans of Earth, then why aren't they candidates for salvation as well?

Interestingly, as a side note to add into this discussion about the applicability of salvation, God felt it necessary to reveal complex oddities such as the Nephilim, who were the heinously evil offspring of fallen angels and humans (Genesis 6:4). Does salvation pertain to them? After all salvation isn't for the angels but applies to humans, so what about a half-angel, half-human hybrid? In order to avoid confusion, God revealed through the prophet Isaiah in Isaiah 26:14 that the Nephilim have no resurrection, and therefore have no hope for salvation. They wouldn't have cared anyway because they were so evil, proud, and rebellious by nature—but God knew of this in advance, and felt compelled to inform Isaiah about it.

Isaiah 26:14*

They are dead, they shall not live; they are deceased, they shall not rise: therefore hast thou visited and destroyed them, and made all their memory to perish.

*This scripture speaks of those who are dead, yet will never rise again. Scripture states in Revelation 20:12–14 that all the dead will be raised for judgment on Judgment Day, yet the dead mentioned in Isaiah 26:14 will not be raised, so they are uniquely different from those who must be judged on Judgment Day.

Revelation 20:12–14

And I saw the dead, small and great, stand before God; and the books were opened: and another book was opened, which is the book of life: and the dead were judged out of those things which were written in the books, according to their works. And the sea gave up the dead who were in it; and death and hell delivered up the dead who were in them: and they were judged every man according to their works. And death and hell were cast into the lake of fire. This is the second death.

I find it profoundly fascinating that scripture goes into such detail, especially concerning the applicability of salvation, which has ramifications that extend to

The Earth Prior to the Creation of Adam and Eve 75

extraterrestrial life in the heavens. Because of this possibility, I believe the second part of Jesus's mission spans even beyond Earth (Acts 2:39; Colossians 1:20; Ephesians 3:10), and I believe Jesus may have been referring to this mission field when he spoke of reaching sheep from another fold (John 10:16).

Acts 2:39 (bold emphasis added)

For the promise is unto you, and to your children, **and to all that are afar off**, even as many as the LORD our God shall call.

Colossians 1:20 (bold emphasis added)

And, having made peace through the blood of his cross, by him to **reconcile all things unto himself**; by him, I say, **whether they be things in earth, or things in heaven**.

Ephesians 3:10–11 (bold emphasis added)

To the intent that now **unto the principalities and powers in heavenly places might be known by the church the manifold wisdom of God**, According to the eternal purpose which he purposed in Christ Jesus our Lord…

John 10:16 (bold emphasis added)

And **other sheep I have, which are not of this fold: them also I must bring**, and they shall hear my voice; and there shall be one fold, and one shepherd.

The eternal purpose of God accomplished through Christ Jesus takes on new meaning when we read Acts 2:39, Colossians 1:20, and Ephesians 3:10 with this perspective. Without understanding how salvation extends beyond Earth, it's difficult to interpret Colossians 1:20, which states that there is life in the heavens that requires reconciliation with God.

In relation to the unique topic of cosmic salvation, I have discovered that Israel in particular stands alone for both the quantity and variety of UFO and alien-related phenomena.[16] Sightings of UFOs in Israel include every size and shape, and aliens range from humans to grays to giants and everything else imaginable. These sightings are sometimes reported not by a few people, but by hundreds of people in mass sightings. Barry Chamish, Israel's leading UFOlogist, has documented a large number of UFO incidents that have occurred in a UFO wave engulfing Israel beginning in 1993.[17] In Barry's book, *Return of the Giants*, Barry describes one such incident that occurred in 1996 that had the government up in arms, thinking they were being invaded by high-tech terrorists. The reported incident included a large gathering of police officers and military personnel that actually fired everything they had, including shoulder-harnessed antiaircraft missiles, at a UFO hovering about one hundred fifty feet above the ground. The UFO just sat there for over two hours, unaffected, then disappeared.[18]

I suspect the reasoning behind the quantity and uniqueness of UFO and alien-related phenomenon in Israel comes from the fact that UFO and alien-related phenomenon *always has* proliferated the region, dating back to biblical times. Concerning beings who are faithful to God, or they are seeking him out, they are here to either help us (Hebrews 1:14) or learn from us (1 Peter 1:12; Ephesians 3:10; 1 Corinthians 4:9). They're searching in Israel for more information about salvation (this is especially the case with those who are mortal like us, searching for a hope of salvation). It could even be that angels are giving tours of Israel to visitors from elsewhere, both learning and teaching the Gospel and fulfilling their task of ministering to the heirs of salvation (Hebrews 1:14; Ephesians 3:10), whoever and wherever those heirs may be.

> 1 Peter 1:12 (bold emphasis added)
>
> Unto whom it was revealed, that not unto themselves, but unto us they did minister the things, which are now reported unto you by them that have preached the gospel unto you with the Holy Ghost sent down from heaven; **which things the angels desire to look into.**
>
> 1 Corinthians 4:9 (bold emphasis added)
>
> For I think that God hath set forth us the apostles last, as it were appointed to death: **for we are made a spectacle unto the world, and to angels, and to men.** We are fools for Christ's sake, but ye are wise in Christ; we are weak, but ye are strong; ye are honorable, but we are despised.

All the activity of the entire world (and even the entire universe), centers on Israel whether the world governments or the Christian church even realize it. God's plan of salvation for humanity (and possibly others—which is what I'm talking about here) has always centered on the Jewish people. It began with the father of the Jews, Abraham; was fulfilled through a Jewish Messiah; was written by Jewish authors in a Jewish book (the Bible); and was spread throughout the entire Earth, initially through twelve Jewish apostles.

Even in the eternal future, God's plan of salvation will be commemorated in Jewish feasts and celebrations forever (feasts of Passover and unleavened bread—Ezekiel 45:21; first fruits—Ezekiel 44:30; Pentecost or Weeks—Ezekiel 46:9; trumpets—Ezekiel 44:5; day of atonement—Ezekiel 45–46; and tabernacles—Ezekiel 45:25; Zechariah 14:16–21). Christianity is actually a Jewish religion, and is only a Gentile (non-Jewish) religion as well because God adopted the Gentile people of the world into the family of God. This family of God started with his chosen people, the Jews (Romans 11:17–23; Matthew 22:2–10; Mark 7:26–29). The people of Earth should be aware of this great truth; this is especially the case for the Christian church. Even a great variety of otherworldly beings coming from light-years away, showing up in vast numbers to the small nation of Israel, know more about the significance of Israel than many people on Earth. Many of these visitors may want to be adopted too!

Now I've rambled on about the fact that Earth is important even from the heavenly perspective, but I still haven't answered why Earth was chosen as the spearhead of God's plan of reconciliation (which I believe applies to both Earth and the heavens); this is what I've been building up to all this time. I apologize for any perceived long-windedness, which is very difficult for me to avoid. Frequently the material used to answer questions raises numerous additional questions in itself—questions I am compelled to explain before continuing. Hopefully my wordiness is not too disjointed to follow.

Continuing on, I have learned that what makes Earth significant in the universe concerns the very pieces of the puzzle to Earth's history that I was missing when I wrote *Aliens in the Bible*. There was not just one catastrophic flood in Earth's history; there were two. Furthermore, the Bible speaks of a civilization that was destroyed in the first flood, prior to Adam and Eve ever being created. Such a civilization implies that the inhabitants who were destroyed were not human. Even more intriguing is the fact that these beings were ruled under the kingdom of Lucifer; he ruled Earth at that time. This is where his connection to Earth was initially established. It's no wonder he's been so intimately tied to this particular planet. Earth is significant not solely because God's love for humanity is any greater than his love for any other intelligent species in the universe, but because Earth is the seat of rebellion in the heavens that started with Lucifer even before humanity was created. That's why Christ came to Earth. It is why Earth is a universal mecca of angelic attention among the vastness of the cosmos, and it is the place where Satan will eventually be isolated, and his rebellion will be crushed once and for all (Revelation 20:9).

But how did I discover all of this? I didn't; Finis Dake, the theologian I mentioned earlier, did nearly twenty years before I was born. Finis Dake taught me about the law of double reference—among many other things, all of which shed light on a number of scriptures that explain everything I've just mentioned. While the Watcher site referenced earlier mentioned much of the same material

I'm about to expound upon here, the research wasn't backed up nearly as well as Finis Dake's. One of the key elements that would've helped the Watcher site greatly is a brief explanation of the law of double reference.

Now I will briefly describe the law of double reference. There are numerous scriptures that speak of a visible person who is immediately addressed, while at the same time, an invisible person who is using the visible person as a tool to hinder the plan of God is also addressed. For example, when Jesus said to Peter, "Get thee behind me, Satan: for thou art an offense to me: for thou savor not the things that be of God, but those that be of men" (Matthew 16:23), he did not mean that Peter was the personal devil, but that Peter was being manipulated by Satan. From Jesus's perspective, he might have even been able to see Satan standing right there, hence both Peter and Satan were addressed and involved in the statement Jesus made. The only way to interpret such passages is to understand that part of the passage is addressed to a man, while the elements that have no relevance to the visible man in question must clearly be referring to the spiritual manipulations working through the man.

Now that it is understood to some degree how the law of double reference works when spotted in scripture, I will now illuminate some Old Testament scriptures, some of which require the law of double reference to interpret. These scriptures speak of a flood predating the flood of Noah, which Finis Dake referred to as the flood of Lucifer.

1. The Flood of Lucifer Mentioned in the Old Testament

The flood of Lucifer is first alluded to in Genesis 1:2, which states the following (bold emphasis added): "The earth was without form and void, and darkness was over the face of the deep. And the Spirit of God was hovering over the face of the **waters**." The wording in Genesis 1:2 is significant because it proves that the original creation of the heavens and Earth was completed in its entirety in Genesis 1:1, and that the re-creation of Earth into a habitable state follows from Genesis 1:2 to Genesis 2:25, as I will now explain.

> Genesis 1:1–2 (bracketed comments added)
>
> In the beginning God created the heaven and the earth [the creation is complete]. And the earth was without form, and void; and darkness was upon the face of the deep [earth was destroyed]. And the Spirit of God moved upon the face of the waters [earth is covered with water].

Like many people, I used to separate Genesis 1:1 from the rest of the account of creation by thinking of it as a qualifying opener. I thought the author opened with a brief one-line summary to serve as an outline, as a form of literary style, then followed with the details. Supporting this position is the fact that the sun, moon, and stars were made on the fourth day of creation. How could the "heaven and the earth" have been created in Genesis 1:1, when we can see that the sun, moon, and stars, which constitute the heavens, were made on the fourth day of creation?

While this question may seem difficult to answer, translating Genesis 1:1 as being nothing more than an outline raises a number of other intriguing questions, which are even more difficult to rectify than the sun, moon, and stars being made on the fourth day of creation.

First, if Genesis 1:1 is an outline of creation, it's not a very good one: it doesn't establish the order of creation correctly. If Earth were made before the heavens, then Genesis 1:1 should state that God made Earth, and then the heavens, rather than the heavens and Earth. Wouldn't the order of these details be somewhat significant, considering the fact that this outline documents the origin of all creation?

Then there's Job 38:4–7, which states that the heavens were in existence before the foundations of Earth were laid. How is this possible, if the heavens weren't even made until the fourth day of creation, which didn't occur until *after* Earth had form, land, and even vegetation?

Job 38:4–7

> Where wast thou when I laid the foundations of the earth? Declare, if thou hast understanding. Who hath laid the measures thereof, if thou know? Or who hath stretched the line upon it? Whereupon are the foundations thereof fastened? Or who laid the corner stone thereof; when the morning stars sang together, and all the sons of God shouted for joy?

Continuing on, I'm certain that God can do anything he wants, and in any order. But I always thought that it was extremely odd that Genesis seems to indicate that Earth was created before the sun around which it rotates. I'm no astrophysicist, but something about that just doesn't seem right.

What about the days of creation? How could there be days if there was no sun until the fourth day? One could postulate that the word "day" symbolizes an unknown period of time, a creative age, such as a thousand years (taken from Psalms 90:4 and 2 Peter 3:8, which stipulate that a day to the Lord is as a thousand years to man). The language of the creation account, however, qualifies days with a beginning and an ending. In every other place in the entire Bible, days qualified in this manner refer to a literal twenty-four-hour period. Interpreting the word "day" in a different way in this part of the Bible makes for a nonstandard method of interpretation; it's an extremely weak argument, and subject to opinions and biases rather than interpretational standards. Even harder to rectify with the idea of a creative age is the fact that vegetation was created on the third day. Exactly how long could the vegetation created on the third day have survived without sunlight? Are we to believe that vegetation grew in the complete absence of sunlight for a thousand years?

I want to know what the Bible actually says—not what I want it to say. And when something doesn't make sense, the incongruence is never a contradiction, but rather a passage highlighted by God for more revelation. I can't count the number of times God has taught me something new in scripture because I was trying to resolve a perceived contradiction. I have discovered that the Bible makes perfect sense, both logically and in consistency, if people would simply study it, believe what it says, and weigh out all the implications of all of its facets.

Concerning these perceived contradictions, of which Genesis had a list I contemplated for years, it is even harder to squeeze the Bible's mention of angels into the creation account when the Bible speaks of angels, yet never mentions when they were created. This is another mystery that bellows in its conspicuous absence. Hidden within this mystery lies a clue to Earth's ancient history. How do those who argue against the interpretation of Genesis 1:2 through Genesis 2:25 as being the account of a re-creation of Earth into a habitable state explain questions that arise concerning angels? Certainly they dwell in the heavens, and the heavens had to be created before the angels, so the angels would have a place to exist—so were angels created on day four, five, or six? If it took an entire day to create Adam, then why wouldn't it take a longer period of time to create all of the angels, who, according to scripture, are even greater than man, both in glory and in number (Psalms 8:5; Hebrews 2:6–9)?

Psalms 8:5

For thou hast made him a little lower than the angels, and hast crowned him with glory and honor.

Hebrews 2:6–9

But one in a certain place testified, saying, What is man, that thou art mindful of him? Or the son of man, that thou visit him? Thou made Him a little lower than the angels; thou crowned Him with glory and honor, and set him over the works of thy hands: Thou hast put all things in subjection under his feet.

> For in that He put all in subjection under Him, He left nothing that is not put under Him. But now we see not yet all things put under him. But we see Jesus, who was made a little lower than the angels for the suffering of death, crowned with glory and honor; that He by the grace of God should taste death for every man.

The existence of angels is first established with Satan's arrival in the Garden of Eden. For humanity it's early in life, seeing that Adam and Eve were innocent, naive, and didn't have any children yet. It's evident that Satan, however, was prowling about much longer than Adam and Eve. In Ezekiel 28:11–19 and Isaiah 14:12–14, we're told that Satan tried to lead an invasion into heaven at one time, yet failed, and his kingdom was destroyed. As I questioned earlier, did he once rule an entire kingdom, deceive his kingdom as well as many other worlds, then orchestrate a massive, interstellar invasion into the highest heaven—in less time than it took Adam and Eve to have their first male child? This has to be the case, if one believes Genesis 1:1 falls within a six-day construct of the heavens and Earth.

As always, whenever there is a perceived incongruence in scripture, the deficiency is due to a lack of human understanding, rather than an inconsistency in scripture. In this case, I was very fortunate to have Finis Dake's thirty years of research on hand, which pointed me in the right direction.

In order to rectify the incongruence of the fourth day of creation with Genesis 1:1, it makes more sense to suggest that the heavens were in fact already created, but not illuminated. The darkness that covered Earth in Genesis 1:2 was removed on the fourth day, revealing the sun, moon, and stars, instead of the heavens being created on the fourth day of creation. In the English translation of the Bible, this isn't particularly evident, but one clue that stands out is a distinction between the purposeful translation of the words "create" and "made."

The word "created" is from the Hebrew word "bara," meaning, "to create," in the sense of making something new or bringing something into existence without the use of pre-existing material. This latter idea is certainly true of the materials out of which the heavens and Earth were formed. In Hebrews 11:3, we read that the "things which are seen [the visible things] were not made of things that do appear" or that were visible. If the heavens and Earth were brought into existence, then it is certain that at one time they were not in existence. "Bara" is found forty-nine times in the Hebrew Bible, and it is translated as "create," eight times; "created," thirty-three times; "make," four times; "Creator," three times; and "createth," one time. The primary idea is to bring into existence something new that never existed before, in any way, shape, or form.

"Bara" is used only seven times in Genesis 1:1–2:4, the passage that records all of the creative ages. It is correctly translated "created" in each case. In all other verses in this passage of scripture, the word "made" is used. It is from the Hebrew word "asah," meaning to make something out of already existing material.

In Genesis 1:1, the heavens and Earth are created, or brought into existence; in Genesis 1:21, the sea creatures are created, or brought into existence; and in Genesis 1:28, man is created, or brought into being. Thus the word "bara" is reserved for the introduction of each of the three great spheres of existence: the world of matter, the world of natural life as in all living creatures on the Earth, and the world of spiritual life on Earth as represented by man. The heavens and Earth were brought into existence "in the beginning," while the living creatures and man were brought into existence on Earth, on the fifth and sixth days of the restoration of Earth to a habitable state. All other accomplishments in the six days were not of a *creative* nature—things were *made* out of already existing material that had been created in the various periods or ages "in the beginning."

The sun, moon, and stars were *made* on the fourth day of creation (or re-creation, to be more precise), as described in Genesis 1:14–19.

> Genesis 1:14–19 (bold emphasis added)
>
> And God said, Let there be lights in the expanse of the heavens to separate the day from the night. And let them be for signs and for seasons, and for days and years, and let them be lights in the expanse of the heavens to give light upon the earth. And it was so. And God **made** the two great lights—the greater light to rule the day and the lesser light to rule the night—and the stars. And God **set** them in the expanse of the heavens to give light on the earth, to rule over the day and over the night, and to separate the light from the darkness. And God saw that it was good. And there was evening and there was morning, the fourth day.

In the English language, everything except the words "made" and "set" allow this to mean that God removed the darkness that covered Earth. This is especially evident where it states that the work of the fourth day was to "give light on the earth." As already previously elaborated, the word "made" does not mean that God created these objects from nothingness; it actually means in this passage that God simply restored Earth so that it received light again. As for the word "set," *Strong's Enhanced Lexicon* reveals that it can be used with "great latitude of application" and can be translated with other words as well, such as appoint, assign, ascribe, charge, commit, apply, add, and so on.[19] God "set" the sun, moon, and stars in the expanse of the heavens to give light on Earth could easily be reworded to say that God "assigned," "ascribed," "committed," or "charged" the sun, moon, and stars in the expanse of the heavens. Such a translation would have a much different connotation in the English language than using the word "set."

The Hebrew word for heaven is "shamayim," meaning lofty, sky, the higher ether where the celestial bodies revolve. It is found over 650 times in the Bible, and in most places, it should have been translated "heavens." God created the heavens, which includes the sun, moon, stars, and all the inhabitants of heaven and all things therein (Nehemiah 9:6). Then he created Earth and its inhabitants and all things therein. After that, an unknown period had transpired when angels ruled the universe. Then something happened on Earth that caused God enough anguish to obliterate it, because in Genesis 1:2 the Earth was void, and without form. This is where the flood of Lucifer comes into the picture.

The translation of the word "was" from the Hebrew "hayah" is the verb "to become," not the verb "to be." It is translated "became" 67 times (Genesis 2:7, 19:26, 20:12, 24:67; Exodus 4:3–4; Numbers 12:10, and many others); "becamest" 2 times (1 Chronicles 7:22; Ezekiel 16:8); "came" and "came to pass" 503 times (Genesis 4:3, 6:1, 6:4, 11:2, 11:5, and many others); "become" 66 times (Genesis 3:22, 18:18, 48:19, and many others); "come" and "come to pass" 131 times (Genesis 4:13, 6:13, 6:18–20, 27:40, and many others); and many times "be" in the sense of "become" (Genesis 1:3, 1:6, 1:9, 1:14, 3:5, and many others).

The words "without form" are from the Hebrew word "tohu," which means waste, desolation, or confusion. It is translated "waste" (Deuteronomy 32:10); "without form" (Genesis 1:2; Jeremiah 4:23); "vain" (Isaiah 45:18; 1 Samuel 12:21); "confusion" (Isaiah 24:10, 34:11, 41:29); "empty" (Job 26:7); "vanity" (Isaiah 40:17, 40:23, 44:9, 59:4); "nothing" (Job 6:18; Isaiah 40:17); and "wilderness" (Job 12:24; Psalms 107:40). Simply using *Strong's Enhanced Lexicon* will reveal all of these translations.[20]

How could Earth *become tohu* if it isn't even created yet? Answer: it was created in Genesis 1:1, and in Genesis 1:2, it was annihilated.

Delving deeper, the Hebrew word for "earth" is "erets," meaning "dry ground." Genesis 1:9–10 states, "And God called the dry land Earth," so we can read Genesis 1:1, "In the beginning God created the heavens and the dry land." In Genesis 1:2, the dry land is then covered with water, proving that it was flooded after its original creation (Psalms 104:6–9; 2 Peter 3:5–6).

Genesis 1:9–10

And God said, Let the waters under the heaven be gathered together unto one place, and let the dry land appear: and it was so. And God called the dry land

Earth; and the gathering together of the waters called he Seas: and God saw that it was good.

Psalms 104:6–9

Thou covered it with the deep as with a garment: the waters stood above the mountains. At thy rebuke they fled; at the voice of thy thunder they hasted away. They go up by the mountains; they go down by the valleys unto the place which thou hast founded for them. Thou hast set a bound that they may not pass over; that they turn not again to cover the earth.

2 Peter 3:5–6

For this they willingly are ignorant of, that by the word of God the heavens were of old, and the earth standing out of the water and in the water: Whereby the world that then was, being overflowed with water, perished…

Thus the heavens and Earth were completed and inhabited, and then Earth was flooded in Genesis 1:2 before the beginning of the six days of Genesis 1:3–2:25.

From these passages, it should be clear what Genesis 1:2 is actually saying. God did not originally create Earth in a state of chaotic desolation. It is stated in Isaiah 45:18 that God did not create Earth *tohu* (vain, or desolate), yet in Genesis 1:2 Earth *became tohu*. If Earth was not originally created desolate, then it must have been created and inhabited, then later *became* desolate.

Isaiah 45:18 (bold emphasis and bracketed comments added)

For thus says the LORD that created the heavens; God himself that formed the earth and made it; He hath established it, **He created [bara] it not in vain, He formed it to be inhabited**: I am the LORD; and there is none else.

Reiterating Genesis 1:1, God first created the heavens and Earth. After that, perhaps several billion years transpired between Genesis 1:1 and Genesis 1:2; scripture states in Job 38:5–7 that there was a period of time when angels dwelled among the stars in harmony with God. Concluding the dispensation of angels, as theologians call it, was Satan's deceptive cavorting around the universe, deceiving the angels into his scheme to invade heaven and overthrow God. It was at this point that Earth was destroyed (Genesis 1:2).

But wait a minute. Why destroy Earth? Did God think, "Gee, Lucifer over here has deceived a bunch of his fellow angels, so I think I'll change his name to Satan, and then go and destroy this obscure little planet on the south side of the universe for no particular reason"? Was destroying Earth God's way of beating a pillow in frustration?

The fact that Earth was destroyed before the creation of Adam and Eve is the first clue indicating that its destruction was linked to Lucifer's fall from glory. But what does Lucifer's rebellion against God have to do with the destruction of Earth? The answer is simple; Earth was his home, and it was destroyed by a worldwide flood that predates the creation of Adam and Eve.

The flood of Lucifer is both an Old Testament and New Testament doctrine. Moses (the author of Genesis, which I already covered), Isaiah, Ezekiel, and Jeremiah listed in the Old Testament, and Jesus, Peter, Paul, and John in the New Testament, all spoke of the flood of Lucifer.

Isaiah 14:12–14 provides the first clear piece of evidence that Lucifer ruled Earth before the days of Adam. This passage covers a period in Lucifer's history when he ruled, and then was defeated, so it had to be a time before Adam was created. It wasn't until after Adam was created and deceived by Lucifer that Lucifer, who was then known as Satan, was able to regain dominion of Earth as the prince of this world, as he is today—but not indefinitely. To understand this passage, the law of double reference comes into play, because the passage alludes to both Satan and the king of Babylon. The king of Babylon is the visible person who is immediately addressed in Isaiah 14:4, and Satan, who is also addressed in Isaiah 14:12–14, is the invisible entity who is using the visible person as a tool to hinder the plan of God.

Isaiah 14:12–14 makes statements that obviously don't apply to the king of Babylon, even though he is the one who is initially addressed. The statements directed toward Satan are those that I will now highlight.

Isaiah 14:12–14 (bold emphasis added)

How art thou fallen from heaven, O Lucifer, son of the morning! How art thou cut down to the ground, which didst **weaken the nations**! For thou hast said in your heart, I will **ascend into heaven**, I will exalt my **throne** above the **stars** of God: I will sit also upon the mount of the congregation, in the **sides of the north**: I will **ascend above the heights of the clouds**; I will be like the most High.

These statements tell us a great deal about Lucifer's fall from glory, which occurred at a time that predates Adam and Eve. First Lucifer was a king, because he had a throne; this also implies that he had subjects over whom he ruled. Isaiah also states that he weakened nations, which had to be composed of other nonhuman living beings (angels and those who had not become angels yet), because

Adam and Eve weren't created yet during the time that Lucifer was a sinless being ruling as a king.

Of most concern to Earth's ancient history is the fact that Lucifer's kingdom was on Earth. If this were not the case, he never could have ascended above the clouds, stars, and into the highest heaven. When a kingdom is located under the clouds, it has to be on a planet, and the Bible only alludes to other planets indirectly as being celestial spheres in the heavens. If Satan were in the heavens already when he began his invasion into the highest heaven where God's throne is, then he would've already been among the stars from Isaiah's perspective, rather than under the clouds. How could he be in the heavens if he was under the clouds? His ascension above the clouds, and then to the stars, naturally implies that his point of origin was Earth.

Initiating his rebellion from Earth, Satan did, in fact, ascend above the clouds, and then out among the stars to other planets. He established dominion throughout the universe, where he deceived one-third of the stars (implying the angels among them—Revelation 12:4). Moreover, Ezekiel 28:11–17 speaks of Lucifer's ability to walk amid the stones of fire (most likely referring to intergalactic travel among solar systems). This was a privilege that Ezekiel states was revoked when Lucifer was defeated after his invasion into the highest heaven.

Isaiah 14:12–14, as well as numerous other key scriptures, plainly reveals that Earth's significance in the universe has to do with its history, and all the clues point to the fall of Lucifer: hell is in *Earth*; Christ came to *Earth* to establish his plan of salvation; and Satan will be confined to and defeated on *Earth*.

Note also from this passage of scripture in Isaiah that the ground, clouds, stars, and heaven were already created, even before Adam and Eve existed. This is further evidence pointing to the account of creation from Genesis 1:2–2:25 being the restoration of Earth to a habitable state, rather than an original creation.

Take further note that God's throne is located in the northern hemisphere of the universe (also mentioned in Psalms 48:2, 75:6–7). Why else would a physical direction be associated with God's throne?

The thrones, dominions, principalities, and powers in heaven and on Earth, referenced in Colossians 1:15–18, are the authorities Satan deceived. They were located throughout the heavens and on Earth. Lucifer was given a kingdom, which he expanded to include those he deceived. From this, as well as from other scriptures that will follow, it appears that his kingdom was centered on Earth. His subjects were earthly creatures, and those who were not angels were destroyed in the flood of Lucifer. Thus it can be argued that Isaiah teaches that Earth was inhabited before Adam, and was ruled by Lucifer, whose kingdom was overthrown when he rebelled.

<u>Colossians 1:15–17</u>

> Who is the image of the invisible God, the firstborn of every creature: For by him were all things created, that are in heaven, and that are in earth, visible and invisible, whether they be thrones, or dominions, or principalities, or powers: all things were created by him, and for him: And he is before all things, and by him all things consist.

I already mentioned Ezekiel 28:11–19, but only touched on one small component of this scripture. It actually deserves its own exposition, just as I've given to Isaiah 14:12–14. This passage of scripture also requires the law of double reference to interpret it. The visible man addressed is the king of Tyrus, while again it is Satan who is addressed as the invisible entity working through the visible man.

<u>Ezekiel 28:11–19</u> (bold emphasis added)

> Son of man, take up a lamentation upon the king of Tyrus, and say unto him, Thus says the Lord GOD; Thou seal up the sum, **full of wisdom, and perfect in beauty**. Thou hast **been in Eden the garden of God**; every precious stone was thy covering, the sardius, topaz, and the diamond, the beryl, the onyx, and the jasper, the sapphire, the emerald, and the carbuncle, and gold: the workmanship of thy tabrets and of thy pipes was prepared in thee in the day that thou wast created. Thou art **the anointed cherub that cover**; and I have set thee so: thou wast **upon the holy mountain of God**; thou hast **walked up and down in the midst of the stones of fire**. Thou wast **perfect in thy ways** from the day that thou wast created, till iniquity was found in thee. By the multitude of thy merchandise they have filled the midst of thee with violence, and thou hast sinned: therefore I will cast thee as profane out of the mountain of God: and **I will destroy thee, O covering cherub, from the midst of the stones of fire**. Your heart was lifted up because of thy beauty; thou hast corrupted thy wisdom by reason of thy brightness: I will cast thee to the ground; **I will lay thee before kings, that they may behold thee**. Thou hast **defiled thy sanctuaries** by the multitude of your iniquities, by the iniquity of thy traffic; therefore will I **bring forth a fire from the midst of thee**, it shall devour thee, and I will bring thee to ashes upon the earth in the sight of all them that behold thee.

For starters it should be obvious that the law of double reference is required to interpret this passage of scripture, because the king of Tyrus had never been in the Garden of Eden, nor was he a cherub that covers. This scripture is clearly speaking about Lucifer. Some of the information divulged about Lucifer is commonly known, such as the fact that the sin that caused him to fall from glory was pride. He was full of himself, amazed at his own beauty, and felt that he deserved much more than what God had granted him. A detail that is commonly overlooked, however, is that Lucifer was found in the Garden of Eden at a time that predates Adam and Eve. This fact should be clear, because the context of the scripture is describing what Lucifer was like before he fell from glory: full of wisdom and perfect in beauty; the anointed cherub that covers [protects]; perfect in his ways. He certainly wasn't perfect in his ways when he deceived Eve into eating the forbidden fruit—so this description of Lucifer's life as a sinless being goes back much further than Adam and Eve. Because of this, we must deduce that the Garden of Eden that Lucifer inhabited was yet another Garden of Eden that was destroyed in the flood of Lucifer in Genesis 1:2.

Why did Lucifer's garden have the same name as Adam and Eve's garden? That's easy to answer—because it's actually the same garden from God's point of view. He simply destroyed it, then re-created it again. The fact that this garden retained its name is yet further evidence that Lucifer once reigned on Earth.

How long was his reign? Who knows? Perhaps it was some 4.5 billion years or so, as some geophysicists speculate to be the age of Earth.[21] What meaning does time have in a perfect society of immortal beings?

Even after Lucifer sinned, a considerable amount of time must have transpired before Lucifer decided to invade God's throne. First we see Lucifer being cast out of the mountain of God, then destroyed from amid the stones of fire, which I elaborated on extensively in my book *Aliens in the Bible*. Lucifer was an incredibly powerful being, so his walking in the midst of the stones of fire is not a summer night's stroll in the garden down a path lined with glowing pebbles. These stones of fire are directly linked with Lucifer's access to the heavens, because he was restricted from them when he fell from glory. Using the language of Ezekiel's day, Ezekiel couldn't have picked clearer terminology to describe solar systems—"stones of fire." A sun is a stone that is on fire, is it not? What could be clearer? It's up to us to conclude that these stones weren't ordinary, but rather formidable enough to be worth mentioning, to say the least.

Ezekiel also mentioned Lucifer's sanctuaries, which Lucifer defiled. This infuriated God to the point of complete indignation. Most likely many of these sanctuaries included other glorified worlds that Lucifer was corrupting during his walks amid the stones of fire. Lucifer's spread of deception throughout the universe is also referred to in Revelation 12:4, which is a prophecy that is still in progress. Lucifer is currently dragging one-third of the stars from the sky (which I interpret to be angels who dwell among the solar systems in the universe). This prophecy will be completely fulfilled in the last days, when these otherworldly beings will flood into Earth.

Certainly it took a great deal of testing before God finally destroyed Earth. I speculated in *Aliens in the Bible* that the "fire from the midst of thee," which God used to destroy Lucifer (Ezekiel 28:18), might actually be a direct reference to the massive planetary explosion I described earlier. If there was a civilization on Astera (or Rahab), all of its mortal inhabitants were obliterated. Fragments of this explosion also resulted in the destruction of the atmosphere on Mars (and any civilization that may have existed there) and the destruction of Lucifer's kingdom on Earth. In essence it was the ultimate wrath of God on the center of Lucifer's kingdom. Earth's entire solar system was thrown into a blender to such a degree that even solar regulation was in chaos, which is what we see in Genesis 1:2.

This incredible demonstration of God's destructive power wasn't carried out on a whim. As previously stated, it may have taken millions, if not billions of years before God finally decided to bring about such destruction. I say this for two reasons: first, God is slow to anger and quick to forgive (Psalms 103:8, 145:8), and second, it's evident in the fossil record. Fossils, by most scientific accounts, are extremely old. Some dinosaurs are estimated at over 245 million years old.[22] The fossil record actually documents five mass extinctions in Earth's ancient history, dating at 440, 345, 245, 195, and 66 million years ago, assuming scientists are accurate about their assessments of the fossil record.[23] Several mass extinctions would not be out-of-synch with God's character documented in scripture. Most likely, these other destructions, which may have included the Ice Age, a collision with a comet, or a massive volcanic explosion, didn't destroy all life on Earth. They were smaller destructions, leading up to the flood of Lucifer, when the last global catastrophe destroyed all life on Earth, leaving no trace of survivors.

I always give more credibility to scripture rather than what scientists think they know, but if the scientists are concluding facts that line up with scripture, then I'm apt to agree with them.

Psalms 103:8

The LORD is merciful and gracious, slow to anger, and plenteous in mercy.

Psalms 145:8

The LORD is gracious, and full of compassion; slow to anger, and of great mercy.

Going back to why I used to believe in the young-Earth theory—I remember that one of the main reasons I believed that Earth was young (about six thousand to seven thousand years old, when Adam and Eve were created),[24] was because dinosaurs were vicious creatures that would not have been compatible with a sinless world devoid of death. Since sin and death entered the world through Adam and Eve, I believed that dinosaurs, or any carnivores for that matter, could not have existed before Adam and Eve. When I learned that sin had entered Earth through Lucifer even before Adam and Eve were created, however, as Ezekiel 28:11–19 clearly outlines when using the law of double reference, I finally understood that there is no reason why dinosaurs couldn't be millions of years old. In essence, it doesn't matter how old they are, because the Bible remains true, whether they're old or young.

Theologians who argue against the *gap theory* (the theory I espouse in this book, which suggests that there was is *gap* of time between Genesis 1:1 and Genesis 1:2), do so because they think that sin existing in the Earth prior to Adam and Eve is in conflict with Genesis 1:31, which states that everything that God made was good. I disagree with such reasoning, because I don't see a conflict. Sin and death existed in the Earth prior to Adam and Eve, until Genesis 1:2, after which the Earth was re-created into a habitable state. It was at this point that God cleansed the Earth, removing the curse of sin and death.

In Genesis 1:31, it states, "And God saw every thing that he had made, and, behold, it was very good..." God's *creation* would not be *very good* if sin existed in the universe, creationists argue, but this assumption is only applicable if the phrase, "everything that he had made," was in reference to the entire universe, in which case, it's not. If the writer of this passage wanted to indicate that *everything that God **created*** was very good, he would have said, "everything that he had **created**, and, behold, it was very good..." The Hebrew word for "made," however, was distinctly chosen by the author, rather than the Hebrew word for "created." I already covered the differences between these two words earlier, so the only thing I wish to point out here is that the Hebrew word for "made" was selected for Genesis 1:31. Therefore, everything that God restored from Genesis 1:2 and onward, was *very good*. This precludes the corruption that existed in the heavens at that time, in places where Satan and his minions dwelt.

I fully understand that creationists who argue against the old-Earth theory do so primarily because some old-Earth advocates compromise scripture by trying to make it jive with the theory of evolution. I agree with this motivation, because I stand ardently against the theory of evolution, but I also think that creationists are barking up the wrong tree. They can argue about the validity of carbon dating all they want, but the root issue is that the Bible is true, and evolution is WRONG (God is the Creator, not random chance), irrespective of whether the Earth is old or young.

Concluding this exposition on Ezekiel: even more clearly than Isaiah, Ezekiel teaches that Lucifer had a kingdom on Earth, and that there were people here before Adam. The evidence doesn't end with Moses, Isaiah, and Ezekiel, however. Jeremiah has a few words to chime in on this matter before leaving the Old Testament, and then there's Peter, Paul, John, and Jesus in the New Testament, who also describe the same events.

In Jeremiah 4:23–27, the prophet Jeremiah describes a vision of Earth after it was destroyed by the flood of Lucifer, in Genesis 1:2. The wording is virtually identical to Genesis 1:2 so that it wouldn't be missed. At the time this vision was given to Jeremiah, the Israelites were far away from God. God therefore gave Jeremiah this vision to instill righteous fear in them by showing them how completely awesome his destructive power was demonstrated at one time. (He also gave him this vision to reveal something about Earth's ancient history.)

> Jeremiah 4:23–27 (bold emphasis added)
>
> I beheld the **earth**, and, lo, it **was without form, and void**; and **the heavens**, and they **had no light**. I beheld the mountains, and, lo, they trembled, and all the hills moved lightly. I beheld, and, lo, **there was no man**, and **all the birds of the heavens were fled**. I beheld, and, lo, **the fruitful place was a wilderness**, and **all the cities thereof were broken down** at the presence of the Lord, and by his fierce anger. For thus hath the Lord said, The whole land shall be desolate; yet I will not make a full end.

To begin with, we can clearly see that two events are being described. The first event is the complete destruction of Earth and all living things; even light was withheld from the heavens. The words "was without form" are the same as those in Genesis 1:2, meaning that Earth was ("hayah"—became) without form ("tohu"—completely desolate and empty). Therefore it is evident that the destruction spoken of is something that had already happened long ago. Later in the passage, however, the Lord states (bold emphasis added), "The whole land **shall be** desolate; yet **I will not make a full end**." This is something that is *going* to happen, and the level of destruction is not as devastating as it was in the past, because it won't be a complete end.

If we key in on the first destructive event that this passage speaks of, several facts are revealed that place this scripture into unity with the other scriptures that also document the flood of Lucifer. First, what time in Earth's history describes a cataclysmic event of the magnitude described by Jeremiah? This passage states that God destroyed every single city that existed on the entire Earth. The only event in Earth's history aside from the flood of Lucifer that comes close to this is the flood of Noah, but all those aboard the ark survived the flood of Noah, while this particular event left no trace of life whatsoever (no men...not even birds). This event also included earthquakes on a global scale (mountains were trembling and hills were moving—all of which crushed a large animal population deep within the Earth, leaving behind much of the fossil record we have today), as well as a darkness that overshadowed the entire Earth; none of these details were documented in the account of the flood of Noah. One could argue that darkness was caused by clouds, but clouds don't completely block out the light of the sun to the degree that the context of this scripture stipulates.

This flood that Jeremiah speaks of can't possibly be the flood of Noah.

A second item of interest in this scripture is the fact that the heavens were in existence, as well as Earth, when this destruction took place. Concerning the heavens, lights from the heavens were withheld from shining on Earth, thereby causing the darkness across the face of the Earth, as spoken of in Genesis 1:2. This darkness wasn't removed until day four of the restoration of Earth into a habitable state. In essence the wording of Jeremiah 4:23–27 not only resembles the wording of Genesis 1:2, it also harmonizes with it perfectly, so that we have clues pointing us back to Genesis 1:2 to help us fill in a few more details that are quietly hiding between Genesis 1:1 and Genesis 1:2.

What are these details that many people today aren't aware of? The biggest detail here is the fact that before Adam and Eve were created, there were men (some of which were translated into angels—note that they used to be men, but we know that at least some of them became angels, because Lucifer was their king). Furthermore there were also cities on Earth at this time predating Adam and Eve. We probably can't imagine how advanced this society was; they had possibly billions of years of uninterrupted progress in which to enjoy their lives and experiment, and seek and discover the mysteries of the cosmos. Even one of these beings would be capable of completely revolutionizing the Earth of today. What would people be like if they lived so long they could obtain a doctorate in every field of study that existed? We simply can't imagine how intelligent these beings really were, and in some cases, still are—concerning those that were angels.

As stated earlier, I knew of the pre-Adamite interpretation of Jeremiah 4:23–27 before I wrote *Aliens in the Bible*; I read about it on the Watcher site. The information I was missing in order to accept it, however, was how this scripture

The Earth Prior to the Creation of Adam and Eve   93

harmonizes with Isaiah and Ezekiel when using the law of double reference. Without the law of double reference, it appeared to be taken out of context, but putting the law together with Genesis, Isaiah, and Ezekiel makes the pieces of the puzzle fit together too perfectly to ignore.

Then there's the New Testament, which also lends support to the flood of Lucifer.

2. The Flood of Lucifer Mentioned in the New Testament

When speaking from the New Testament, who better to start with than Jesus Christ? What did Jesus have to say about the flood of Lucifer?

In Matthew 13:35, Jesus used a curious expression: "from the foundation of the world," which lost a vital component of its meaning in the King James translation. Further investigation into this phrase actually reveals that it should have been translated "from the overthrow of the social world."

> Matthew 13:35 (bold emphasis added)
>
> That it might be fulfilled which was spoken by the prophet, saying, I will open my mouth in parables; I will utter things which have been kept secret **from the foundation of the world**.

The Greek word for "world," being "kosmos," means social system or social world, order, or arrangement. As for the Greek word for "foundation," it has two meanings in Greek. The first is the noun, "themelios," and the verb related to it, "themelioo." These are ordinary words used for the foundation of a literal building, or of an organization, or of the work of a person. Places where this word was translated as such are found in Matthew 7:25, Luke 6:48–49, 1 Corinthians 3:11, Ephesians 2:20, and Hebrews 6:1.

> Matthew 7:25 (bold emphasis added)
>
> And the rain descended, and the floods came, and the winds blew, and beat upon that house; and it fell not: for it was **founded** upon a rock.

> Luke 6:48–49
>
> He is like a man which built an house, and dug deep, and laid the **foundation** on a rock: and when the flood arose, the stream beat vehemently upon that house, and could not shake it: for it was **founded** upon a rock. But he that

hears and doeth not, is like a man that without a **foundation** built a house upon the earth; against which the stream did beat vehemently, and immediately it fell; and the ruin of that house was great.

1 Corinthians 3:11

For other **foundation** can no man lay than that is laid, which is Jesus Christ.

Ephesians 2:20 (bold emphasis added)

And are built upon the **foundation** of the apostles and prophets, Jesus Christ himself being the chief corner stone...

Hebrews 6:1 (bold emphasis added)

Therefore leaving the principles of the doctrine of Christ, let us go on unto perfection; not laying again the **foundation** of repentance from dead works, and of faith toward God...

The second interpretation of this word is the noun "katabole" and its corresponding verb, "katabollo," which are not the ordinary words for foundation, as they are sometimes translated in the New Testament. While *Strong's Enhanced Lexicon* offers the less common definition of "katabole", listed as G2602, it begins with a direct reference to its corresponding verb, G2598 for "katabollo."[25] Looking up the word "katabollo" in *Strong's Enhanced Lexicon* reveals that it actually means "to throw down" or "cast down," and it is translated with this meaning in 2 Corinthians 4:9, and Revelation 12:10.

2 Corinthians 4:9 (bold emphasis added)

Persecuted, but not forsaken; **cast down**, but not destroyed...

Revelation 12:10

And I heard a loud voice saying in heaven, Now is come salvation, and strength, and the kingdom of our God, and the power of his Christ: for the accuser of our brethren is **cast down**, which accused them before our God day and night.

The actual meanings of these two words in the original Greek are pivotal to understanding many passages in the New Testament that refer to the flood of Lucifer. In all occurrences except Hebrews 11:11, the word "katabole," which I already explained is a variation of "katabollo," is the word used in the Greek language, and it's directly connected with the Greek word "kosmos," referring to an ancient social order that was destroyed at a time predating human history. In *all* instances, the phrase "foundation of the world" should have been translated as "overthrow (or disruption, or ruin) of the social world," but this translation was specifically avoided. Why was it avoided?

The translators correctly understood this phrase to refer to a time that marks the beginning of the world (as re-created in Genesis 1:2 through Genesis 2:25), rather than to the flood of Noah. This understanding was derived from the fact that the expressions before and since the disruption of the world in scripture refer to God loving his Son, of choosing men in Christ, and of Christ being foreordained "before the disruption of the world" (John 17:24; Ephesians 1:4; 1 Peter 1:19–20).

> John 17:24 (bold emphasis added)
>
> Father, I will that they also, whom thou hast given me, be with me where I am; that they may behold my glory, which thou hast given me: for thou loved me before the **foundation of the world**.
>
> Ephesians 1:4 (bold emphasis added)
>
> According as He hath chosen us in him before the **foundation of the world**, that we should be holy and without blame before him in love…
>
> 1 Peter 1:19–20 (bold emphasis added)
>
> But with the precious blood of Christ, as of a lamb without blemish and without spot: Who verily was foreordained before the **foundation of the world**, but was manifest in these last times for you…

Scripture also speaks of things kept secret, of God planning a kingdom, of prophets being slain, of a book of life being prepared, of the world of the six days of re-creation, and of Christ suffering in death only once, since the disruption of the world (Matthew 13:35, 25:34; Luke 11:50; Hebrews 4:3, 9:26; Revelation 13:8, 17:8). All of these scriptures clearly point to an eternal past rather than to the flood of Noah, because God has always loved his Son, he always knew which men he was going to choose in Christ, Christ was always foreordained for the mission he undertook—and the list goes on. It would make no sense to make such statement using the flood of Noah as a line of demarcation, yet the phrase "disruption of the social world," would easily be mistaken as the flood of Noah by many who read it. Because of this, the translation became "foundation of the world," which bears with it a much simpler explanation, rather than "overthrow of the social world," even though the actual translation doesn't match up exactly with what the Greek words really intended.

So the translation "foundation of the world" has always been clearly distinguished from the flood of Noah (in the English language), as it should be, but the fact that there have been two worldwide destructions (documented in the Bible) in Earth's history, rather than one, was greatly obscured in this distinguishing process. This was a most unfortunate side effect of the English translation that was chosen.

Fortunately, scripture is highly redundant concerning important events, and this property has saved the original translation for us today. Any truth that can be found in scripture can usually be arrived at by a number of different scriptures that all point to the same thing from multiple angles. Evidence of the flood of Lucifer is infused throughout both the New and Old Testaments, so this translation has not been lost.

Further adding to the preservation of scripture is the fact that the Greek and Hebrew translations are all available for free to anyone who wants to download them from the Internet. My personal recommendation is the Bible software program I previously mentioned, called e-Sword (http://www.e-sword.net/bibles.html).[26] E-Sword includes Greek and Hebrew references, lexicons, topical guides, and a number of other resources, all for free. With tools like e-Sword, researchers can simply look up a word, such as "foundation," and see for themselves the things I'm talking about here. I challenge anyone reading this who doubts me to use this program. Look up Matthew 13:35, 25:34, Luke 11:50; John 17:24; Ephesians 1:4; Hebrews 4:3, 9:26; 1 Peter 1:20; and Revelation 13:8, 17:8. All these scriptures speak of a "disruption of the social world," rather than the "foundation of the world."

Christ himself taught, as recorded by the Holy Spirit through Matthew, Luke, and John, that there was a disruption or ruin of the social order on Earth at a point in time that marks the beginning of humanity in this current age. He mentioned such a doctrine four times while on Earth (Matthew 13:35, 25:34; Luke 11:50; John 17:24), and twice to John on Patmos after he had ascended to heaven (Revelation 13:8, 17:8). All of these references were obscured in translation, because the translators were trying to distinguish this event as the beginning of the creation (or more accurately, the re-creation) of Earth, rather than to the flood of Noah. Not all was lost, however; there were enough clues left behind in order for the people of today to put the pieces of this ancient puzzle back together.

Aside from Jesus's use of the phrase "disruption of the social world," the apostle Peter also used phrases distinguishing the line of demarcation that was drawn when the world was destroyed in the flood of Lucifer. In 2 Peter 3:6–7, Peter referred to the world that existed before the flood of Lucifer as "the world that then was." The Greek word for "world" is "kosmos," which, as previously stated, means social system. The social system "that then was" is specifically differentiated from the social system of today. Why did Peter do this?

2 Peter 3:6–7 (bold emphasis added)

Whereby **the world that then was**, being **overflowed with water**, perished: But the **heavens and the earth**, which are **now**, by the same word are kept in store, reserved unto fire against the day of judgment and perdition of ungodly men.

First, the social system that we have today is the entire species of human beings on Earth. If the social system that is *now* were the same one as the one that *then was*, there would be no point in distinguishing it from the one that then was. So what distinguishes these social systems?

For the sake of argument, one could postulate that the social system before the days of Noah was simply different from the social system of today, but this suggestion is weak. The flood of Noah left survivors; the social system that is *now* actually comes from the social system that was before the flood of Noah, so it's the same social system. Nothing intrinsic has changed about it. We're all human beings that came from Adam—so the fundamental aspect of our social system has not changed.

Clinching the argument that there is a fundamental difference between today's social system and the social system Peter was speaking of, however, are the further details expressed in 2 Peter 3:6–7, where Peter elaborates with the reference to "the heavens and the earth, which are now." One might be able to argue that the social system of Earth has changed since the flood of Noah, though this is a weak argument. But certainly the heavens weren't changed in the flood of Noah. Noah's flood only affected Earth. The flood spoken of in 2 Peter 3:6–7, however, affected even the heavens. This is the same flood mentioned in Genesis 1:2, where it speaks of a global catastrophe so severe that included, in addition to a flood completely covering Earth that left no survivors, God's obscuring all light from shining on Earth. The entire solar system was in utter chaos up until day four of creation, when God restored solar regulation (stabilizing the orbits of all the celestial bodies like planets, moons, and comets), and reestablished cosmic order to mark signs, days, years, and seasons (Genesis 1:14). The heavens that were then, it may therefore be surmised, were obliterated, as would be the case in the event of an explosion of a large planet sending fragments railing throughout the solar system, destroying every trace of life on all the planets in the entire solar system.

The scriptures of 2 Peter 3:6–7 and Genesis 1:2 harmonize perfectly and reconcile all the discrepancies of elements that speak of a massive flood from antiquity that isn't the flood of Noah, because it's a flood that predates the beginning of the account of creation (or restoration of the Earth to a habitable state). Understanding this flood also reconciles Genesis 1:14–18 with Job 38:5–7; the events of day four of creation must be understood as the restoration of solar regulation, because Job 38:5–7 states that the morning stars (Hebrew "kokab"—literal meaning, "stars") sang together when the foundations of Earth were being laid. (The singing obviously refers to spiritual harmony among the inhabitants of the heavens, and the literal singing of the angels that inhabited the stars.)

Taking all this into account, I consider Peter's words alluding to the world *that then was* that was destroyed by a flood, to be among those that teach about the flood of Lucifer.

So far I've mentioned what Jesus and Peter said about the flood of Lucifer, but the apostles Paul and John spoke of this flood as well.

Like Jesus, Paul speaks of the foundation of the world (overthrow of the social order), referring to a time before Adam (Ephesians 1:4; Hebrews 4:3; 9:26).

Ephesians 1:4 (bold emphasis added)

According as He hath chosen us in him before the **foundation of the world**, that we should be holy and without blame before him in love…

Hebrews 4:3 (bold emphasis added)

For we which have believed do enter into rest, as He said, As I have sworn in my wrath, if they shall enter into my rest: although the works were finished from the **foundation of the world**.

Hebrews 9:26 (bold emphasis added)

For then must He often have suffered since the **foundation of the world**: but now once in the end of the world hath He appeared to put away sin by the sacrifice of Himself.

There's no need to go into great depth about Paul's teaching on this subject, because his teaching is nearly identical to Jesus's teaching. The same Greek words that Jesus used, Paul uses exactly the same way, referring to a period that predates

Adam and Eve. Briefly put, according to Paul, God's plan of redemption outlined in Ephesians 1:1–14 (which includes a massive abundance of wonderful blessings), was established before the overthrow of the social order.

As for the apostle John, he too used the same exact phrase "foundation of the world" (overthrow of the social order) in conjunction with a time frame that predates Adam and Eve. Revelation 13:8, 17:8 states that the book of life was prepared since the foundation of the world. Obviously, the book of life was prepared long before the flood of Noah, so the disruption of the social order that John refers to clearly predates Adam and Eve.

> Revelation 13:8 (bold emphasis added)
>
> And all that dwell upon the earth shall worship him, whose names are not written in the book of life of the Lamb slain from the **foundation of the world**.
>
> Revelation 17:8 (bold emphasis added)
>
> The beast that thou sawest was, and is not; and shall ascend out of the bottomless pit, and go into perdition: and they that dwell on the earth shall wonder, whose names were not written in the book of life from the **foundation of the world**, when they behold the beast that was, and is not, and yet is.

The book that I obtained the majority of my research from concerning the flood of Lucifer, *God's Plan for Man,* goes into far more detail than I have covered here. Finis Dake's extensive comparative analysis chart comparing the flood of Lucifer with the flood of Noah is an exceptional compilation of facts that is very difficult to refute. In summation, anyone wanting more scriptural support concerning the flood of Lucifer can simply obtain *God's Plan for Man*. I feel confident that the elements I have extracted from Dake's research here are sufficient, however, to prove my point that there was, in fact, a flood that destroyed an ancient kingdom on Earth, which was ruled by Lucifer, long before Adam and Eve were ever created.

One last note to point out concerning the flood of Lucifer before I move on: many people of the past have most likely stumbled upon this information before, but willingly disregarded it. This continues to happen today, too, because 2 Peter 3:5–7, which immediately precedes Peter's statements about the world that then was, states that in the last days, people will be willingly ignorant of this flood.

<u>2 Peter 3:5–7</u> (bold emphasis added)

For this **they willingly are ignorant** of, that by the word of God the heavens were of old, and the earth standing out of the water and in the water: **Whereby the world that then was, being overflowed with water, perished**: But the heavens and the earth, which are now, by the same word are kept in store, reserved unto fire against the day of judgment and perdition of ungodly men.

This leads me to believe that information concerning this flood contains a somewhat weighty degree of importance. Interestingly, the Greek word translated as "ignorant" in 2 Peter 3:5 and 3:8 is "lathano," which is translated elsewhere as "hid" and "hidden" (Mark 7:24; Luke 8:47; Acts 26:26).

<u>Mark 7:24</u> (bold emphasis added)

And from thence He arose, and went into the borders of Tyre and Sidon, and entered into an house, and would have no man know it: but He could not be **hid**.

<u>Luke 8:47</u> (bold emphasis added)

And when the woman saw that she was not **hid**, she came trembling, and falling down before him, she declared unto him before all the people for what cause she had touched him, and how she was healed immediately.

<u>Acts 26:26</u> (bold emphasis added)

For the king knows of these things, before whom also I speak freely: for I am persuaded that none of these things are **hidden** from him; for this thing was not done in a corner.

It's as if Satan has obscured this truth; he doesn't want the people of Earth to know about the humiliating destruction of his kingdom that occurred so long ago. Concerning this flood, there may also be other things that are hiding, and knowing about this flood will flush them out. Perhaps part of his grand strategy of deception that the Antichrist will present to the world will not be as successful as he intends if more people are knowledgeable concerning these facts.

Imagine open contact with extraterrestrial life from other worlds becoming a reality. Before anyone has any time to react, these beings could possibly introduce themselves by helping us uncover archeological discoveries that prove the existence of a thriving, intelligent, technologically advanced civilization that once populated Earth an extremely long time ago. What better way for them to introduce themselves than to say they've been here before, and that we either have descended from them or were genetically engineered by them? There may be physical evidence available, and they might educate us with accurate details about much of Earth's ancient history, but their interpretation of all they reveal will be a deception. What might the people of Earth make of such evidence? Who could argue against this physical proof?

Could it be that such archeological discoveries have already been made, and we don't even know what they are yet?

3. Who Were the Cro-Magnon and Neanderthal?

Returning to the previously mentioned fossil record, biologists believe that modern humans evolved from earlier ape-like hominids, dating back to Australopithecus Africanus, some 2.5 million years ago.[27] According to evolutionary theory, one species evolved into another, until homo sapiens evolved about 250,000 years ago.[28] Within the homo sapiens classification are Neanderthal (*homo sapiens neanderthalenis* to be more precise), and the more recent Cro-Magnon. Both Neanderthal and Cro-Magnon had a larger brain capacity than the humans of today.[29] This fact doesn't fit with most creationist theories, *or* evolution. (Creationists that believe Neanderthal were the same as the humans of today have to explain the difference in cranial capacity, which is a bone structure not likely to be altered through any known natural process if excluding evolution. Evolutionists, on the other hand, must wonder why cranial capacity suddenly changed course, since having a larger cranial capacity was clearly the dominant trait up until this point.) While the Cro-Magnon were almost identical to modern humans, the Neanderthal were shorter, stockier, and had thicker bone density.[30]

Bringing this into a biblical perspective, how does one understand these early homo sapiens from scripture? If Adam and Eve were created only 6000 to 7000 years ago, how could these *humanoids* exist 250,000 years ago?

The typical creationist theory suggests that the results of carbon dating have been dramatically skewed because of the flood of Noah.[31] Furthermore, modern human fossils are found all over the globe that are carbon dated to millions of years old. One example is a modern human arm fragment that was found in Kenya that was carbon dated to 4.5 million years old.[32] How could modern humans evolve from the ape-like creature Australopithecus Africanus if they predate them?

While these arguments raise doubts as to the current theory of evolution and the credibility of carbon dating, they still don't explain everything. The Kenya finding (as well as many others I haven't mentioned), might make it possible to prove that Australopithecus Africanus isn't a common ancestor to modern humans, but it still doesn't explain the existence of humans that are not descended from Adam and Eve. If human extraterrestrials reveal themselves in the future, Christians will be forced to conclude that not all humanoid-type species are descended from Adam and Eve. This will put them back to square one with most of their theories about pre-Adamites (namely, the predominant theory that they didn't exist). In light of this, let's assume for a moment that Neanderthal, Cro-Magnon, and earlier hominids, such as homo erectus and homo habilis, existed before Adam and Eve. Might these pre-Adamite humanoids be members of Lucifer's ancient kingdom of the past? Furthermore, if they were part of Lucifer's kingdom, were they truly human, or were they something else?

Various authors have espoused the existence of pre-Adamites, yet I've only found the work of Finis Dake (and those similar to him) to remain consistent with scripture, while all other pre-Adamite theories appear to be riddled with inconsistencies.

Author Russell Grigg, in his article titled *Pre-Adamite man: were there human beings on Earth before Adam?* posted on www.answersingenesis.org, mentions several pre-Adamite theories, all of which are in error.[33] He refutes the existence of pre-Adamites, as well as the flood of Lucifer, just as I did at one time. I would still refute the pre-Adamite theory as he does if I were looking at the same research that he discusses in his article. Suffice it to say, the information I have in this book about pre-Adamites is nothing like any of the theories mentioned in Grigg's article.

Concerning the pre-Adamite theories discussed by Grigg, back in 1655, a Frenchman by the name of Isaac La Peyrère departed from the Bible's clearly stated words when he suggested that Adam was not created from the dust of the Earth. Instead, Peyrère published his theory that Adam, Cain's wife, and the inhabitants of Cain's city, all came from pre-Adamite stock. This later opened the door for others to adopt a classification system concerning who was truly *human*, and who was *less than human*. The result was polygenism, which gave rise to a justification for racism, and ultimately slavery.

On a lighter note, leading progressive creationist Hugh Ross explains a new pre-Adamite perspective for this modern generation, which is not tarnished with polygenism (but it's still not right). He defines pre-Adamites as spiritless creatures with human-like characteristics. I'm not sure why he thinks they didn't have spirits. Humans aren't the only beings the Bible mentions as having spirits.

Any theory about pre-Adamites that fails to mention the flood of Lucifer must explain the means by which original sin entered the Earth prior to Adam and Eve, because the pre-Adamites were mortal (at least some of them, because we have their bones). Hugh Ross fails on this critical point. Without explaining how sin entered the Earth prior to Adam and Eve, the existence of pre-Adamites becomes unbiblical. Theories that support doctrines that veer in the direction of evolution, rather than remain consistent with the Bible, might as well not include the Bible at all.

Now that a brief history of flawed pre-Adamite theories has been disclosed, I will divulge what I believe is the correct version of who the pre-Adamites really were. I will further explain how their existence actually harmonizes with the Bible, rather than disputing it.

To begin with, Satan's kingdom was destroyed; a kingdom, by definition, has subjects. Who were these subjects? Were they *angels*? Certainly some of them were, because Satan was an angel. Perhaps I should ask the question, Were *all of them* angels? What happened to these beings, anyway? How could they be destroyed if they were angels, since angels are supposed to be immortal?

3.1. Three Types of Flesh

To explain the identity of the pre-Adamites, it must first be understood that the Bible describes beings composed of three types of flesh. The primary factor differentiating the three types of flesh is the level of mortality.

The first type of flesh is glorified and immortal—so long as the being in question eats from the tree of life, or some form of this tree. Adam and Eve had this form of flesh when they were first created; scripture documents specifically that the way that they physically died was through the blocking of their access to the tree of life (Genesis 3:22–24).

> Genesis 3:22–24
>
> And the LORD God said, Behold, the man is become as one of us, to know good and evil: and now, lest he put forth his hand, and take also of the tree of life, and eat, and live for ever: Therefore the LORD God sent him forth from the Garden of Eden, to till the ground from whence he was taken. So He drove out the man; and He placed at the east of the Garden of Eden Cherubims, and a flaming sword which turned every way, to keep the way of the tree of life.

Adam and Eve were not angels; they were married to each other and sexually reproduced according to God's command in Genesis 1:22 where he told them to "Be fruitful, and multiply" (which is something Jesus said that angels don't do—Matthew 22:30; Mark 12:25; Luke 20:35). Adam and Eve were still immortal, as the angels are, but not to the point that dying isn't even possible. Their immortality was only secure as long as they ate from the tree of life; angels, however, don't require eating anything at all (which I will soon elaborate on). Had Adam and Eve never eaten from the tree of the knowledge of good and evil, they would never have died (Genesis 3:22–24—the tree of life would have enabled them to continue to live forever, which is why they were restricted from it). Eventually they would've been translated into angelic form, as Jesus said will eventually happen to all believers, and they wouldn't need to eat from the tree of life anymore.

This is what I call the butterfly theory, which documents the origin of angels. Life-forms created in God's image begin as pseudoimmortal, reproductive beings, and then God translates them into an angelic, nonreproductive form. Most likely not everyone is translated at the same time; hence the necessity of the tree of life for all future generations, even in the eternal future of humanity, because these trees are found in New Jerusalem in the future, and will be there forever (Revelation 7:9; 22:1–3).

> Revelation 7:9 (bold emphasis added)
>
> After this I beheld, and, lo, a great multitude, which no man could number, of **all nations, and kindreds, and people**, and tongues, stood before the throne, and before the Lamb, clothed with white robes, **and palms in their hands**...
>
> Revelation 22:1–2 (bold emphasis added)
>
> And he showed me a pure river of water of life, clear as crystal, proceeding out of the throne of God and of the Lamb. **In the midst of the street of it, and on either side of the river, was there the tree of life, which bare twelve manner of fruits, and yielded her fruit every month: and the leaves of the tree were for the healing of the nations**. And there shall be no more curse: but the throne of God and of the Lamb shall be in it; and his servants shall serve him...

The tree of life is in New Jerusalem not for those translated into angels (because angels are irrevocably immortal and don't need the tree of life—Luke 20:36). Instead the tree is present for those who survive the battle of Armageddon, and also those who are born in the new kingdom (life spans will be prolonged, but not necessarily made endless as with the immortal angels—Isaiah 65:20; Zechariah 8:4; Luke 1:33). So many inhabitants of the new kingdom will depend on the tree of life to sustain them until their translation into the highest angelic form.

This first category of flesh requires the tree of life in order to stay alive. This flesh is also subject to damage and diseases; beings in this category can even be killed. Consider the fact that the leaves of the tree of life are for the healing of the nations (Revelation 22:2). Healing would be irrelevant if there wasn't a need. In short, Adam and Eve were probably extremely resilient to disease and physical damage, with miraculous regenerative properties in their flesh. They would have never aged or died from any form of natural cause, but I suspect it was possible to injure them or even kill them (though God could just bring them back to life, as in fact he will).

The second type of flesh is cursed and mortal, though it contains within it an immortal spiritual being incapable of dying or being completely destroyed (*the spirit is not corruptible*—1 Peter 3:4), hence the reason that the destruction of the lake of fire lasts forever.

1 Peter 3:4

> But let it be the hidden man of the heart, in that which is not corruptible, even the ornament of a meek and quiet spirit, which is in the sight of God of great price.

Cursed flesh that houses the spirit of a nonbeliever will eventually be cast into the lake of fire on Judgment Day, whereas cursed flesh that houses the spirit of a believer in Christ will one day be redeemed, so it is not without value. God wouldn't bother to physically resurrect (and completely physically reconstruct people from ash or dust in most cases), if it wasn't possible to undo the curse and glorify the flesh to a state of even higher glory than that of Adam and Eve, which I will now discuss.

The third type of flesh is angelic flesh. There are three characteristics about angelic beings that distinguish them from all other life-forms. First, they are immortal without any dependence on the tree of life; second, contrary to common belief, they do have a type of physical body; and third, they do not reproduce (if they want to remain faithful to God, though they are capable of it). I've already discussed this information to some extent, but now I will elaborate with more detail about the characteristics of angelic flesh.

When I wrote *Aliens in the Bible*, I thought that if an angel sinned, it might become mortal and eventually physically die. My continued Bible research, however, has led me to discover that I was in error concerning this matter. Angels are immortal to the degree that they are incapable of dying, because the Bible clearly states this in Luke 20:36. Therefore angels do not depend on eating from the tree of life. The fallen angels who are chained in hell (2 Peter 2:4; Jude 1–6) and in the Euphrates River (Revelation 9:14) are still alive and physically trapped, even though they could not have eaten from the tree of life in thousands of years.

Luke 20:36 (bold emphasis added)

Neither can they die any more: for they are **equal unto the angels**; and are the children of God, being the children of the resurrection.

2 Peter 2:4 (bold emphasis added)

For if God spared not the angels that sinned, but cast them down to hell, and **delivered them into chains** of darkness, to be reserved unto judgment…

Jude 1:6 (bold emphasis added)

And the angels who kept not their first estate, but left their own habitation, He hath reserved in **everlasting chains** under darkness unto the judgment of the great day.

Revelation 9:14 (bold emphasis added)

Saying to the sixth angel which had the trumpet, Loose the four **angels which are bound** in the great river Euphrates.

The fact that angels are immortal without any dependency on anything is the reason the fallen angels were cast out of the highest heaven and chained up, rather than killed. I haven't found a single instance of scripture speaking of human *spirits* being chained in hell, but fallen angels who have been apprehended are *always* chained.

The only way to completely destroy fallen angels is to throw them into the lake of fire for all eternity, which will eventually happen—but not until after God allows them to run their course, purging from creation (with their deceptions) the aspect of free will that leads to sin, death, and ultimate destruction.

In the future, when humans are translated into angelic form to be just like their angelic brothers and sisters, they too will no longer need to eat from the tree of life, and 1 Corinthians 6:13 states why: people will eventually not need to eat at all (though they will be capable of it). According to 1 Corinthians 6:13, both the stomach and food will eventually be done away with. The Greek word for "destroy" is "katargeo," which means "made useless," so this transformation is one of the effects of the translation process.

1 Corinthians 6:13 (bold emphasis and bracketed comments added)

> Meats for the belly, and the belly for meats: but **God shall destroy [make useless] both it and them**. Now the body is not for fornication, but for the Lord; and the Lord for the body.

In essence, when people are translated into angelic form, eating will be an enjoyable pastime (angels ate with Lot—Genesis 19:3, food will be eaten in heaven—Luke 22:16, 22:18, 22:30; Revelation 2:7, 2:17; 22:1–2; Psalms 78:25; Exodus 24:11), but it won't be a requirement.

Concerning the physical aspect of angels, take note of the previously mentioned chains. If fallen angels didn't have bodies composed of some type of matter, why would chains be required to bind them? The only angels that are nonphysical are deceased Christians currently awaiting reunification with their physical bodies, and this is only a temporary situation.

Many people have the concept of the spiritual realm backward. They think that angels are nonphysical beings living in a kind of formless consciousness, having the ability to become physical when they want to. According to scripture, however, exactly the opposite is true: angels are physical beings living in physical worlds, yet they have the ability to exhibit characteristics that transcend the laws of physics (such as making themselves invisible—Numbers 22:35; and flying—Genesis 28:12; John 1:51; Revelation 14:6–11, and many others) because of their powers, not because they are nonphysical.

Numbers 22:31

> Then the LORD opened the eyes of Balaam, and he saw the angel of the LORD standing in the way, and his sword drawn in his hand: and he bowed down his head, and fell flat on his face.

Genesis 28:12

> And he dreamed, and behold a ladder set up on the earth, and the top of it reached to heaven: and behold the angels of God ascending and descending on it.

<u>John 1:51</u>

And He said unto him, Verily, verily, I say unto you, hereafter ye shall see heaven open, and the angels of God ascending and descending upon the Son of man.

While scripture documents angels as having both physical and supernatural abilities, their habitation is clearly spelled out in Romans 1:20. We may understand the unseen realms through our understanding of the realms we see. Because of this, the unseen realms are the same as the Earth we know, in the sense of being physical worlds. As I have already explained, many scriptures document heaven as a place containing natural terrain. The heavens are real, physical places, and the beings who dwell in them have physical bodies of some sort. If this were not true, there would be no point in a physical resurrection of the dead.

The heavens are physical, yet separated from each other by a dimensional variance, similar to the way frequencies divide radio stations from each other. Angels are capable of tuning the dial of those frequencies within their bodies to traverse different realms. Without understanding this concept, it's easy to think of angels, as well as all things referred to as spiritual, as being intangible. There must be a distinction, however, to spiritual things that are spiritual in an abstract sense, while other spiritual things are clearly described in scripture with no abstractions at all. The river of life, for example, is certainly a spiritual river, and people can talk about it in a spiritually abstract sense and be completely accurate—but that doesn't negate the fact that it is a real, physical river, in a very nonabstract sense.

Some Christians may contend that angels do not have physical bodies at all, because Ephesians 6:12 states that Christians are not at war with flesh and blood, but with powers, principalities, and spiritual forces in high places (angels), hence the conclusion that angels are not composed of flesh and blood. Such an interpretation can't mean that angels don't have physical bodies, though, and following are two reasons why this is the case.

First, if angels could have bodies of flesh and blood even temporarily, Ephesians 6:12 would contradict numerous other scriptures. The Bible documents numerous examples of the fact that angels can do anything people can do: eat (Genesis 18:8, 19:3; Psalms 78:25), sexually reproduce (Genesis 6), and generally operate in the physical realm of Earth as normal people (Genesis 18:1–19:24, 22:11; 2 Samuel 24; 2 Kings 19:35; Psalms 78:49, 91:11; Matthew 28:2; Acts 10, 12, and many others). All of these scriptures document angels performing tasks that require a physical body of some sort, even if only a temporary one. In light of this, in order to have a noncontradictory interpretation of scripture, Ephesians 6:12 must mean that there is a stark contrast to war against beings of *mortal* flesh and blood as compared to angelic beings. What it can't mean is that angels don't have any kind of physical body, because they clearly do, with hands, feet, eyes, voices, heads, mouths, hair, faces, and other parts—all clearly described in scripture (Daniel 10:5–21; Revelation 8:1–9:21, 14:6–11, 15:1–16:21; Acts 1:10, and countless other scriptures).

> Daniel 10:5 (bold emphasis added)
>
> Then I lifted up mine eyes, and looked, and **behold a certain man** clothed in linen, whose **loins** were girded with fine gold of Uphaz: His **body** also was like the beryl, and **his face** as the appearance of lightning, and **his eyes** as lamps of fire, and **his arms and his feet** like in color to polished brass, and **the voice** of his words like the voice of a multitude.
>
> Acts 1:9–11 (bold emphasis added)
>
> And while they looked steadfastly toward heaven as He went up, behold, **two men** stood by them in white apparel; which also said, Ye men of Galilee, why stand ye gazing up into heaven? This same Jesus, who is taken up from you into heaven, shall so come in like manner as ye have seen him go into heaven.

Second, Jesus said that believers in Christ would one day be like the angels, which are in heaven. The believers Jesus spoke of will have physical bodies that will be resurrected from the dead (1 Thessalonians 4:16). Therefore the angels, who these believers will be like, must also have physical bodies.

> 1 Thessalonians 4:16 (bold emphasis added)
>
> For the Lord Himself shall descend from heaven with a shout, with the voice of the archangel, and with the trump of God: and **the dead in Christ shall rise first…**

3.2. The Inhabitants of Lucifer's Kingdom of the Past

Now that it is understood that there are three types of flesh, pertaining to beings in three different stages of glory (1. glorified and pseudoimmortal—depending on the tree of life; 2. cursed and mortal; 3. angelic—completely immortal), the questions above concerning the inhabitants of Lucifer's kingdom of the past can now be answered. Lucifer's kingdom of the past originally had only glorified beings within it, some of which were in the first stage of existence, as Adam and Eve were—waiting to be translated—and some of which were angels.

When Lucifer sinned and fell from glory, those who were not angels yet were cursed and made mortal, and then eventually died or were killed in the flood of Lucifer. Furthermore, because Lucifer's kingdom was a mixed environment containing both fallen angels and mortals, there were probably Nephilim as well.

How long it took before God destroyed Lucifer's kingdom since the time Lucifer first sinned and unleashed the curse of sin and death on Earth is unknown. What is known, however, is the fact that the mortal beings who were killed in the flood of Lucifer are most likely what we know of as the Cro-Magnon, Neanderthal, homo erectus, and homo habilis. All descriptions of humanoid angels in the Bible describe them as men (three examples among many others: Genesis 18:20–22; Isaiah 14:16; Acts 1:9–11), so it makes sense that they were humanoids of some type, since their appearance is very similar to that of humans today, and they were actually capable of interbreeding with people (Genesis 6).

> Genesis 18:20–22 (bold emphasis and bracketed comments added)
>
> And the LORD said, Because the cry of Sodom and Gomorrah is great, and because their sin is very grievous; I will go down now, and see whether they have done altogether according to the cry of it, which is come unto me; and if not, I will know. And **the men** [angels talking to Abraham] turned their faces from thence, and went toward Sodom: but Abraham stood yet before the LORD.
>
> Isaiah 14:16 (bold emphasis and bracketed comments added)
>
> They that see thee shall narrowly look upon thee, and consider thee, saying, Is this **the man** [speaking of Satan], that made the earth to tremble, that did shake kingdoms?
>
> Acts 1:9–11 (bold emphasis added)
>
> And while they looked steadfastly toward heaven as He went up, behold, **two men** stood by them in white apparel; which also said, Ye men of Galilee, why stand ye gazing up into heaven? This same Jesus, who is taken up from you into heaven, shall so come in like manner as ye have seen him go into heaven.

Biologically speaking, Neanderthal and Cro-Magnon are classified as homo sapiens (though Neanderthal is distinguished with the slight variation of *homo sapiens neanderthalenis*), the same classification as modern humans. This classification is derived from the evolutionary perspective, however, which suggests that they mutated into the humans of today. They were very similar to the humans of today, there's no disputing that, but scripture states that **Adam was created from the dust of the Earth**, not mutated from a predecessor. Therefore if the fossils of Cro-Magnon, Neanderthal, homo erectus, and homo habilis are as old as scientists think they are, then they have to be members of a pre-Adamite civilization. In any case, there is a completely biblical explanation for their existence, no matter how old they are.

Those who were angels already were not killed, because the only way to destroy them is to cast them into the lake of fire. But something was done to prevent them from having the unlimited access to the glorified realms that they originally inhabited (Ezekiel 28:16—recall the previously mentioned stones of fire). Their kingdom was destroyed, and their access to the throne of God became limited to occasions of special invitation.

Now the question remains: Since their home was destroyed, where did they spend all their time when they weren't in the highest heaven?

We know from Genesis 3 that Satan was allowed into the Garden of Eden, where he tempted Adam and Eve through his spiritual manipulation of a serpent. From this fact, it may be assessed that the fallen angels have been allowed access into glorified realms of the second heaven. Revelation 12:9 documents that they (the fallen angels) haven't been cast out of the second heaven yet because Satan still accuses the saints before God day and night (and God's throne is in the third heaven, which is higher in glory than the second heaven). Their direct interaction, at least on Earth, however, is currently restrained by the prayers of the church (2 Thessalonians 2:6–8). Because they can't take over the world by force, as they tried to do in the days of Noah (Genesis 6), they exert their influence on Earth in a hidden manner, through the power of deception. For the most part, the people of Earth are consciously unaware of their presence, though knowledge of these beings proliferate every culture in the world.

Since Satan was able to regain power on Earth after deceiving Adam and Eve, it's likely that he was able to regain similar positions of power in other places throughout the cosmos (in the heavens) using the same tactic he used on Adam and Eve. This appears to be the indication of Revelation 12:4. Those species who are deceived by him are placed under his authority to a certain extent. With these details in perspective, it can be determined that fallen angels are still spread throughout the cosmos. These evil beings are planning and plotting, as before, how they will attempt to overthrow God, yet many if not all of them will eventually be confined to Earth (Isaiah 24:21–23; Revelation 12:4). All their activities will climax in a final confrontation where they will be overpowered and destroyed by God.

> Isaiah 24:21-24 (bold emphasis added)
>
> And it shall come to pass in that day, that **the LORD shall punish the host of the high ones that are on high, and the kings of the earth upon the earth. And they shall be gathered together**, as prisoners are gathered in the pit, and shall be shut up in the prison, and after many days shall they be visited. Then the moon shall be confounded, and the sun ashamed, when the LORD of hosts shall reign in Mount Zion, and in Jerusalem, and before his ancients gloriously.
>
> Revelation 12:4* (bold emphasis added)
>
> **And his tail drew the third part of the stars of heaven, and did cast them to the earth**: and the dragon stood before the woman who was ready to be delivered, for to devour her child as soon as it was born.
>
> *The stars represent the fallen angels that rebelled against God when they aligned themselves with Satan. This scripture reveals that Satan will lead them (cast them down) to the Earth in the future.*

3.3. A New Theory About the Cro-Magnon and Neanderthal

Orthodontist Jack Cuozzo, author of *Buried Alive: the Startling Truth about Neanderthal Man*, recently compiled an amazing collection of facts about Neanderthals.[35]

An accomplished orthodontist of thirty years, Jack Cuozzo first started studying human skeletal structures when he came to the realization that his clients appeared to be maturing younger and younger with each passing generation. His quest to understand this riddle led him to study human skulls of earlier generations in order to compare the bone development. One of the primary means of assessing the age of a person using human skulls concerns a unique fact about the human skull in particular; the human skull continues to grow throughout the entire lifespan. The two areas of the human skull where this is most prominent are the protrusion of the jaw and the thickening of the brow.

Eventually Dr. Cuozzo ended up studying Neanderthal skulls, attempting to assess the ages of Neanderthal children. Using an age-progression analysis, Dr. Cuozzo discovered that Neanderthal skulls belonging to children looked exactly the way normal modern homo sapiens skulls would look if they matured very slowly (with the exception of average cranial capacity).

Putting this into the context of scripture, the Bible documents the average lifespan of human beings as they went through a dramatic change following the flood of Noah. Prior to the flood of Noah, people lived to be hundreds of years old, and a few even reached slightly over one thousand years old (Genesis 5). Each generation following the flood of Noah experienced an increase of degradation of the curse of death, resulting in increasingly shorter life spans. In view of this fact, the assumption of Dr. Cuozzo is that Neanderthals are most likely exactly like the humans of today, except that they were intrinsically different with respect to the aging process and lifespan. The genetic differences in their DNA were the natural result of a long life span, rather than being genetically different at birth. I have a different conclusion, however.

Tying Dr. Cuozzo's research together with what I described above about Cro-Magnon, Neanderthal, homo erectus and homo habilis, the inhabitants of Lucifer's kingdom of the past might have mirrored the same transitory period of time as the generations of humans following Adam and Eve did, but with one major exception. Since many of the inhabitants of Lucifer's kingdom were angels, they must have dwelled in a sinless environment for perhaps thousands, millions, or even billions of years before they sinned. This is much unlike the story with Adam and Eve, who sinned very early on, before any humans were translated into angels. Lucifer's kingdom was very diverse, and probably had very complex components of society where the angels dwelt, while at the same time there were primitive people living in caves.

Archeological evidence reveals many examples of primitive people, which is common knowledge.[34] Each age of human development is neatly outlined in numerous texts, beginning with the stone age in 70,000 BC, the development of farming around 8,000 BC, the development of writing around 3,500 BC, and so forth.[35] As for ancient societies that were technologically advanced, I have recently discovered that there are many archeological discoveries of this as well, though the findings aren't as widely spread or accepted, simply because they don't fit! Consider the following three examples below:

> Over the last few decades, miners in South Africa have been digging up mysterious metal spheres. Origin unknown, these spheres measure approximately an inch or so in diameter, and some are etched with three parallel grooves running around the equator. Two types of spheres have been found: one is composed of a solid bluish metal with flecks of white; the other is hollowed out and filled with a spongy white substance. The kicker is that the rock in which

they were found is Precambrian, and dated to 2.8 billion years old! Who made them and for what purpose is unknown.[36]

Scientists of Bashkir State University have found indisputable proof of an ancient highly developed civilization's existence. The question is about a great plate found in 1999, with a picture of the region done according to an unknown technology. This is a real relief map. Today's military has almost similar maps. The map contains civil engineering works: a system of channels with a length of about 12,000 km, weirs, and powerful dams. Not far from the channels, diamond-shaped grounds are shown, whose destination is unknown. The map also contains some inscriptions. Even numerous inscriptions. At first, the scientists thought it was Old Chinese language. It turned out that the inscriptions were done in a hieroglyphic-syllabic language of unknown origin. The scientists never managed to read it…"The more I learn, the more I understand that I know nothing," the doctor of physical and mathematical science, professor of Bashkir State University, Alexander Chuvyrov admits. "It should be noticed," the professor said, "that the relief has not been manually made by an ancient stonecutter. It is simply impossible. It is obvious that the stone was machined." X-ray photographs confirmed that the slab was of artificial origin and has been made with some precision tools.[37]

A mortar and pestle were found in the stratigraphy at Table Mountain, dated from 33.2 to 55 million years old.[38]

Pictures of the above listed items as well as many others I didn't mention can be viewed on the Internet. To see these pictures, check the referenced Web sites in this book for a start. Furthermore, Internet searches investigating a new field of study called OOPARTS (Out Of Place ARTifactS), will reveal a plethora of information.[39] Extremely unusual artifacts and fossils have been discovered all over the world. In many cases, information about these artifacts is deliberately obscured. The Internet is an excellent tool for researching OOPARTS, and I also recommend reading *Unlocking the Mysteries of Creation*, as well as *Forbidden Archeology*, by Michael A. Cremo and Richard L. Thompson; these books are full of examples like those listed above.[40]

To avoid being quoted out of context for recommending *Forbidden Archeology*, I will point out that Torchlight Publishing, which is dedicated to promoting India's Vedic culture and religious beliefs, is the publisher of *Forbidden Archeology*. The archeological information in the book is a good reference, but in no way will I ever espouse the conclusions the authors of *Forbidden Archeology* derive from those facts. Vedic teachings go into depth concerning the events of Genesis 6, and on a historical note, they contain a degree of accuracy (as do the writings of ancient Sumerians, Greeks, Egyptians, Assyrians, Chinese, and Babylonians), yet they support an unbiblical, polytheistic perspective of those events. God destroyed the world in the flood of Noah because people practiced the beliefs of what Vedic teachings continue to uphold today.

3.4. Does the Bible Give Credence to Atlantis?

A similar situation of mortal and immortal beings living on Earth might have occurred on Earth even after the creation of Adam and Eve. I already spoke briefly about how Genesis 6 documents the escapades of fallen angels interbreeding with humans. What I didn't mention, however, was the possibility that fallen angels not only visited the Earth during the days of Noah, but that they actually cohabitated on Earth with humans. Stories of an ancient kingdom known as Atlantis, which was destroyed in a massive cataclysm (possibly the flood of Noah, or perhaps a volcanic eruption slightly before the flood of Noah), could have been an outpost for fallen angels.

The reason I state that Atlantis may have existed *after* the creation of Adam and Eve but before the flood of Noah is because of the striking similarities between Plato's account of Atlantis, and Genesis 6. Following is an excerpt of Plato's exact words:

> When the divine portion began to fade away, and became diluted too often and too much with the mortal admixture, and the human nature gained the upper hand, they then, being unable to bear their fortune, behaved in an unseemly manner, and to him who had an eye to see, grew visibly debased, for they were losing the fairest of their precious gifts; but to those who had no eye to see the true happiness, they appeared glorious and blessed at the very time when they were full of avarice and unrighteous power.[41]

In short, according to Plato's words, Atlantis was destroyed because divine beings (fallen angels) became diluted with mortals (taking wives and bearing children with them). Shortly after this union, the entire society became corrupt, just as it says in Genesis 6 when "the wickedness of man was great in the earth, and that every imagination of the thoughts of his heart was only evil continually."

While the idea of Atlantis is speculative, this kingdom might have been a very large, yet isolated area where angels dwelled (most likely fallen angels, because Earth was ordained by God to be a human domain, and because Plato's account describes events indicative of fallen angels). The interactions between fallen angels and humans documented in scripture may have been simple visits between these immortal beings living in their Earth base, and mortal humans.

The existence of Atlantis has been debated for over two thousand years. Many people would like to classify Plato's works as pure fiction, but the fact that other ancient Greek works, such as Homer's Iliad, gave rise to the discovery of the lost cities of Troy and Mycenae, which were just as mythical as Atlantis, can't be easily dismissed.[42]

As of this date, numerous findings proliferate the bottom of the Atlantic and Mediterranean, which have no concrete answer.[43] Concerning Atlantis, I find it intriguing that Plato specifically mentioned that its inhabitants flew in *sky chariots*.[44] Does this not sound descriptive of otherworldly beings?

In my book *Aliens in the Bible*, I spoke of the interaction between humans and otherworldly beings spanning the cosmos, but now that I understand the perspective of an ancient Earth with both mortal and immortal inhabitants, this interaction may have been with native, immortal beings as well. Their influence is evident in nearly every ancient culture throughout the world. Even today fallen angels are native to Earth, though not completely embedded in our current dimension, as they may have been before (unless they're underwater, which raises the vast topic of Unidentified Submersible Objects, the aquatic equivalent to UFOs). As the Bible testifies, however, they now inhabit the higher realm known as the first heaven, which is in the outer atmosphere of Earth. Some are also trapped in the lower realm of the Abyss, which I will discuss later.

The realm of faithful and fallen angels in the first heaven can be pictured as an invisible sphere around Earth, which on a higher level of glory makes Earth a much larger planet—something perhaps two and one-half times larger than the Sun. I derive this assessment from the fact that Mount Zion, which will be in the center of New Jerusalem on Earth in the future (Hebrews 12:22–23; Revelation 14:1), will be 1,500 miles high. Revelation 21:16 describes the height as twelve thousand furlongs, which equates to fifteen hundred miles (a furlong is one-eighth of a mile).[45] The presence of such a mountain on Earth means that Earth will have to be much larger than it currently is. Presently, the world's tallest mountain is Mount Everest, and it's only 5.5 miles high (29,035 feet).[46] Therefore Mount Zion will be 273 times higher (1,500/5.5 = 273) than Mount Everest! If the Earth must be 273 times larger than it currently is in order to accommodate a mountain 273 times higher than Mount Everest, then Earth will have to be more than twice the size of the Sun. (The Sun is 109 times larger than the Earth.)[47] Of course, the Sun might be larger than the dimension of the first heaven as well, which explains the increase of light on Earth mentioned in Isaiah (Isaiah 30:26; 60:18–22). I will cover this concept of Earth's future proportions more in depth when I discuss the integration of Earth into the first heaven, near the end of this book.

3.5. The Duration of Lucifer's Kingdom

Scripture gives no direct reference concerning the period between when Lucifer initially sinned and when God finally destroyed his kingdom, but there are indications that it took a very long time. It's likely that thousands, if not millions or even billions, of years of mercy transpired before God put an end to Lucifer's kingdom. The existence of an entire global civilization with nations (Isaiah 14:12) and sanctuaries (Lucifer defiled his sanctuaries—Ezekiel 28:18), indicate an extremely prolonged period. Lucifer's dominion and level of access to the heavens (Ezekiel 28:14—"stones of fire"), where he spent a great deal of time spreading his corruptive influence to lead an attack into the highest heaven, further indicate an exceptionally long period of time. Also, recall that I mentioned earlier that biologists believe the fossil record indicates five mass extinctions in Earth's ancient history. These could have been judgments from God during a time of prolonged mercy, until God finally put a complete end to Lucifer's kingdom.

If the development of the pre-Adamite society followed a course similar to the descendants of Adam and Eve, the fossil remains of Neanderthal and Cro-Magnon (as well as earlier hominids, such as homo erectus, and homo habilis) may look like they are different species of humans. All these humanoid beings are most likely the same species, however (except for Australopithecus Africanus, which looks more like an ape than a human), with the exception of the differences in the aging process. In any case, they're all from a different order of creation than Adam and Eve, but very similar to modern humans in appearance and in genetic structure.

The earliest humanoids, who look the least human, such as homo erectus and homo habilis, were those who lived at the time when the curse of death was initially executed, so they matured very slowly, living to hundreds or perhaps even several thousand years old before they died—hence the dramatic physiological changes. For all we know, it could have taken them even longer to die than Adam and Eve; perhaps they ate from the tree of life for a long time, which gave them an extensive longevity and resulted in massive mutations by the time they finally died.

Most creationist scientists explain the differences of bone structure in homo erectus, homo habilis, and Neanderthal with a variety of theories, including various diseases such as arthritis, rickets, and so forth,[48] but I find Dr. Cuzzo's research to have the most merit, with only one exception, which I've already stated. Neanderthal really weren't descendants of Adam and Eve. Simply looking at the size of their cranium reveals that it's the same size as Cro-Magnon, rather than modern homo sapiens. Modern homo sapiens came from Adam and Eve, whereas all the earlier hominids were pre-Adamites.

An intriguing mystery that has puzzled evolutionists for many years concerns the sudden disappearance of Neanderthal. As soon as Cro-Magnon appeared on the scene, they somehow *replaced* Neanderthal in an extremely short period of time. Evolutionists speculate that the Cro-Magnon either killed all of the Neanderthal, or out-classed them in a competition for limited resources.[49] To date, there is really no definitive scientific explanation for their sudden disappearance, because there's no evidence suggesting the Cro-Magnon were a bunch of murdering thugs, or that they were any more skilled than Neanderthal in hunting and gathering food. Neanderthal simply vanished, and no one knows why. This shouldn't be a mystery to anyone reading this research, however. The Neanderthal suddenly disappeared because the life span of the pre-Adamites was dramatically shortened in one generation, just like what happened with Noah's generation. When God determined the age of humans to be 120 years after the flood of Noah (Genesis 6:3), I believe he was operating under a former precedent set during the reign of Lucifer's kingdom on Earth.

As for the Cro-Magnon, their generations were probably born thousands, or even millions of years after the shortening of the life span. They matured to adulthood at an accelerated rate compared to Neanderthal and earlier hominids, which resulted in their different skeletal structure. The cranial capacity remained the same, yet the overall skeletal structure was altered considerably. Cro-Magnon died very young, compared to their Neanderthal counterparts. The result of this scenario is exactly what we see in the fossil record; two categories of early homo sapiens that are more similar to each other (in respect to cranial capacity) than that of modern homo sapiens. Their existence overlaps only briefly, then suddenly the one that is older (more mutated) suddenly disappears, as if vanishing in only a few generations.

3.6. Deceptions Concerning the Cro-Magnon and Neanderthal

Scientists currently have the bones of ancient humanoid fossils on display in museums throughout the world, which they tout as evidence to support evolution, but Christians currently disagree with this assessment. While their disagreement with evolution is correct, their argument refuting evolution is too simplistic in my opinion. Most Christians think that Cro-Magnon, Neanderthal, homo erectus, homo habilis, and the humans of today are all the same, and the variations between them are no different from the variations seen in the humans of today. As previously stated, they speak of these early humans as having rickets (a deficiency of vitamin D), or other diseases, or they point out the poor quality of the scientific methods and researched data of the constituents they argue against.[50] This logic, will be crushed, however, if faced with the evidence that may become known in the future.

How many Christians today believe that Cro-Magnon fossils are really the bones of a completely different lineage of humanoid destroyed in a flood predating Adam and Eve? Even if fossils aren't as old as scientists think, how will Christians react when they actually come face-to-face with Cro-Magnons who are still alive, exiting from their intergalactic vehicles? What will they say then? Simply knowing such beings exist will be confusing enough, but hearing what they have to say, and being presented with scientifically accurate, physical proof of their existence prior to Adam and Eve, will be difficult for many Christians to deal with.

Such physical proof might include a vast archive of technology buried within the crust of Earth, but who has a biblical explanation for it? As stated previously, many puzzling archeological findings have already been discovered, but I feel certain that much more is to come. In the future, Christians may be forced to answer a number of puzzling questions, and in some cases, scrap practically everything they've understood about the heavens and Earth's ancient history when faced with the deceptions that are on the way. Whatever those deceptions are, they must be mind-boggling, considering that they will have global ramifications affecting every religion in the world, as scripture states. Would not the conclusions I'm drawing here fall into this category?

When extraterrestrials publicly reveal themselves, their version of the story behind their origins will be part of their deception (concerning those aligned with Satan). But Christians who know about what I'm writing here will be better prepared to counter those deceptions. With the information I propose, it doesn't matter how old the Earth is, or how old Neanderthal fossils are; they're still not descendants of Adam and Eve, they have a completely biblical explanation, and it has nothing to do with evolution.

But even without me writing this, I'm actually quite confident that any true Christian won't fall for any deception of the Antichrist or his otherworldly allies, because we all know who Christ is, and no one is able to pluck us from his hand (John 10:25–29). He's not an alien, merely another created being among countless others. He's God, the Creator (Colossians 1:16; Ephesians 3:9). He also won't let any future deception deceive the very elect (Matthew 24:24).

John 10:25–29

Jesus answered them, I told you, and ye believed not: the works that I do in my Father's name, they bear witness of me. But ye believe not, because ye are not of my sheep, as I said unto you. My sheep hear my voice, and I know them, and they follow me: And I give unto them eternal life; and they shall never perish, neither shall any man pluck them out of my hand. My Father, which gave them me, is greater than all; and no man is able to pluck them out of my Father's hand.

Colossians 1:16

For by him were all things created, that are in heaven, and that are in earth, visible and invisible, whether they be thrones, or dominions, or principalities, or powers: all things were created by him, and for him...

Ephesians 3:9

And to make all men see what is the fellowship of the mystery, which from the beginning of the world hath been hid in God, who created all things by Jesus Christ...

Matthew 24:24* (Bold emphasis added)

For there shall arise false Christs, and false prophets, and shall show great signs and wonders; insomuch that, **if it were possible**, they shall deceive the very elect.

If it were possible to deceive the elect, the phrase "if it were possible" would not be stated in this passage of scripture. Part of what defines the elect is the fact that it is not possible to deceive them to the degree indicated in this passage.

How will God counter these deceptions in the future to help his elect? I'm sure his plan is comprehensive and profound far beyond my imagination, but I can list at least two things he'll do as part of his plan. First, he will bring an increase in miracles unlike ever before (John 14:12). Second, his faithful angels will be fitting into this picture, bringing the proper perspective of the heavens, and preaching the Gospel. Don't forget about them! Revelation 14:6–11 states that God's faithful angels will be flying in the air, preaching the Gospel to the people of Earth. (This most likely means not all the spacecraft in the sky will be bad. As for church services, they may turn out to be something quite unusual by today's standards.)

John 14:12 (bold emphasis added)

Verily, verily, I say unto you, He that believeth on me, the works that I do shall he do also; and **greater works than these shall he do**; because I go unto my Father.

Revelation 14:6–11 (bold emphasis added)

And I saw **another angel fly** in the midst of heaven, **having the everlasting gospel to preach unto them that dwell on the earth**, and to every nation, and kindred, and tongue, and people, saying with a loud voice, Fear God, and give glory to him; for the hour of his judgment is come: and worship him that made heaven, and earth, and the sea, and the fountains of waters. **And there followed another angel**, saying, Babylon is fallen, is fallen, that great city, because she made all nations drink of the wine of the wrath of her fornication. **And the third angel** followed them, saying with a loud voice, If any man worship the beast and his image, and receive his mark in his forehead, or in his hand, the same shall drink of the wine of the wrath of God, which is poured out without mixture into the cup of his indignation; and he shall be tormented with fire and brimstone in the presence of the holy angels, and in the presence of the Lamb: And the smoke of their torment will ascend up for ever and ever: and they have no rest day nor night, who worship the beast and his image, and whosoever receives the mark of his name.

So the delusion will be strong (2 Thessalonians 2:11), but God will not allow it to deceive anyone with a heart he's interested in and has chosen (Matthew 22:14). Actually God is the one who is allowing this deception to happen in order to weed out the nominal Christians (Christians by claim only, but not in reality). Those who fall away never had true faith to begin with. Part of the work of this book is to reach out to those inactive, lukewarm Christians sitting on the fence. They're neither dead nor alive. They're frequently mistaken for being nominal Christians (because God is probably the only one who can clearly distinguish them). The events about to overshadow Earth will hopefully wake them up and pull them into a true, strong relationship with God—most likely after the rapture of the church. And this book has its place, among the miracles and the empowerment of the church to come, and the angels preaching the Gospel, all working together to counteract the deceptions that are on the way.

2 Thessalonians 2:11–14

And for this cause God shall send them strong delusion, that they should believe a lie: That they all might be damned who believed not the truth, but had pleasure in unrighteousness. But we are bound to give thanks always to God for you, brethren beloved of the Lord, because God hath from the beginning chosen you to salvation through sanctification of the Spirit and belief of the truth: Where unto He called you by our gospel, to the obtaining of the glory of our Lord Jesus Christ.

Even if I'm wrong about the theories I propose in this book, it doesn't hurt to explore such possibilities. If extraterrestrials never come, then that's just fine! This book will simply be a logical exploration into the possibility and implications of the existence of extraterrestrial life, and nothing more. If, however, they do come, I think it would be a good thing for as many people as possible to have this book on their shelves or stored in their computers. I've done what I can to get the word out; all my books are free in electronic format on my Web sites, http://www.AliensAndTheAntichrist.com, http://www.AliensInTheBible.com, and http://www.EaglestarProphecy.com, and I've chosen the smallest royalty my publisher will allow in order to make the printed version of this book as cheap as possible. If you, the reader, have downloaded this book to your computer, feel free to e-mail it to whomever you like. Knowledge of these things should be common, and I don't want money to be a hindrance.

My audience is currently small, but as I already stated, I suspect that it will one day increase, most likely after I'm gone. Those who thought they were Christians, or used to know Christians (because all the true Christians will be gone after the rapture of the church), will be looking for books called *Aliens in the Bible*, and *Aliens and the Antichrist* when they wake up one day and discover the sky proliferated with alien spacecraft. Having this material in advance will make this transition a smoother one when it happens.

4. Lucifer's Dominion beyond the Earth Before and After Adam and Eve

I've already covered how Lucifer ruled over an angelic population covering Earth, but I haven't gone into great depth about how his kingdom extended throughout the solar system and even into the far reaches of the cosmos. This was obviously the case, since he will successfully deceive one-third of the inhabitants of all of the stars before he's done (Revelation 12:4). Many of these beings united with him a long time ago, when he led an invasion into the highest heaven (Isaiah 14:13), and they will unite with him again in the near future, when he invades Earth in full force through the Antichrist.

After Lucifer was defeated in his attempted invasion, he was finally restrained from having unlimited access to glorified realms (Ezekiel 28:16—"stones of fire"). Earth's entire solar system was obliterated at that time.

Then, some time later, God eventually reestablished solar regulation for Earth's solar system during the restoration of Earth to a habitable state. At the same time, God also most likely constructed a new multidimensional aspect to the entire universe (or heavens). This is what I believe is the primary difference between the heavens that are *now* from the heavens that *were then*, as spoken of in 2 Peter 3:6–7. The time frame for this activity makes sense for two reasons. First, the purpose of a dimensional variance is to prevent cursed beings from accessing and corrupting higher realms; in a sense, it's a spiritual quarantine. Because there were no cursed beings before Lucifer's sin, a dimensional variance wasn't needed until after the rebellion.

Second, hell most likely wasn't constructed until after God put an end to Lucifer's initial rebellion. The construction of hell required a dimensional variance within Earth. (If you skipped to this section, you need to drop back to the onion theory section to understand the multidimensional structure of the universe more thoroughly.)

Even though Lucifer (now Satan) was defeated, his kingdom destroyed, his unlimited access to glorified realms removed, scripture makes it clear that he's definitely not chained up in hell (Tartarus) with many of the other fallen angels who played a part in his rebellion. He and some of his followers were still allowed access to God's throne on occasion, in order to attend specific counsels (Job 1:6–12). It was most likely such a counsel as the one mentioned in the book of Job, where Satan was given access into the Garden of Eden in order to attempt to deceive Adam and Eve. God specifically authorized this event in order that Adam and Eve would be tested, to see if they would use their freedom of choice to obey God, rather than listen to Satan. Satan's access to Eden would have abruptly ended for the last time if either Adam or Eve rebuked him, but we all know how the story went.

Once Satan achieved this victory in the Garden of Eden, his level of access on Earth was increased once again, and his kingdom was reestablished in the first heaven, which surrounds the upper atmosphere of Earth. If this were not the case, he wouldn't be called the prince of the power of the air (Ephesians 2:2), and the god of this world (2 Corinthians 4:4). In addition, other fallen angels who are aligned with him wouldn't be able to rule over entire regions of Earth with such power that they are able to fight against God's faithful angels conducting operations for God. This was exactly the case with the archangel Gabriel, when he was detained by the prince of Persia for twenty-one days before the archangel Michael arrived and helped him to penetrate the first heaven in order to deliver a prophetic message to Daniel (Daniel 10:12–13). The archangel Gabriel stated plainly to Daniel that a second fight would take place when he had to leave, noting the fact that he was going back through the upper atmosphere (first heaven) to leave Earth, where the battle would take place (Daniel 10:20).

> Ephesians 2:2 (bold emphasis added)
>
> Wherein in time past ye walked according to the course of this world, according to the **prince of the power of the air**, the spirit that now works in the children of disobedience:
>
> 2 Corinthians 4:4 (bold emphasis added)
>
> In whom the **god of this world** hath blinded the minds of them which believe not, lest the light of the glorious gospel of Christ, who is the image of God, should shine unto them.
>
> Daniel 10:12–13 (bold emphasis added)
>
> Then said he unto me, Fear not, Daniel: for from the first day that thou didst set your heart to understand and to chasten thyself before thy God, thy words were heard, and I am come for thy words. But **the prince of the kingdom of Persia withstood me one and twenty days: but, lo, Michael, one of the chief princes, came to help me; and I remained there with the kings of Persia.**
>
> Daniel 10:20
>
> Then said he, Do you know wherefore I come unto thee? **And now will I return to fight with the prince of Persia: and when I am gone forth, lo, the prince of Grecia shall come.**

Therefore we know that Satan's dominion was reestablished on Earth, though with certain limitations, and from these facts, it can be deduced that Satan most likely used similar tactics in order to gain footholds elsewhere in the cosmos, as I mentioned earlier. His time is running out, though, and his friends are running out as well. His past is catching up with him, closing in on all sides. He's doing his best to rally his allies, however, because he knows his time is short. The ultimate showdown is about to commence.

PART III
THE NEAR FUTURE

At this current point in time, nothing of monolithic, earth-shattering, reality-twisting proportions is currently happening on Earth. In fact, the last massive earth-shattering event to grace the globe was the arrival of Jesus. Since he was here, the only things of monumental importance (from the heavenly perspective) that have occurred had to do with him in one way or another. Other than that, everything has been business as usual, at least from the perspective of the universe. Many people can actually live their lives thinking that this small, simple life is all there is, and there's nothing up in their faces demanding their attention and stating otherwise, for the most part. This is especially the case in first world countries where religious persecution isn't the norm.

The western mind is especially dull to the realm of the eternal. The only exception to this dullness, interestingly, lies in the imagination of science fiction writers, who bear peculiar insight into the greater reality much more than people realize (which I will soon briefly elaborate on).

A subtle shift in the collective consciousness of the people of the entire Earth is slowly transpiring, however. New Agers can feel it, though they incorrectly interpret it. Christians feel it as well, though their picture is somewhat antiquated (seen through the eyes of nontechnological people who are missing the pieces of the reality which we now grasp, yet do not realize it).

The rate of this change has accelerated with considerable momentum, especially within the past one hundred years. Many people are looking to the stars with a different perspective. A sense of being on the edge of something big is brimming within the hearts and minds of many people. Furthermore a conditioning has been preparing people, especially within the context of a plethora of exceptionally enlightened science fiction authors who are unknowingly seeing into the future to some extent, as I previously stated. The metaphysical concept of alternate realities is clear in the minds of many people, and the prospect of life in outer space is widely accepted, though frequently ridiculed on the surface. Simply consider the popularity of movies such as *Star Wars* and epic television shows such as *Star Trek*, both of which bear a strikingly accurate depiction of the heavens.

People are waiting, anticipating something big, but they're not sure what it is. When it comes, in the blink of an eye, the shockwave of that initial, explosive event will reverberate across Earth in an unprecedented epiphany, though many will not be able to explain what it all means.

4

Could the Rapture of the Church Be a Mass Alien Abduction?

Could it be that the rapture of the church is a bit different from how most Christians imagine it?

If everything I've been writing about in this book concerning faithful and fallen angels, and the host of heaven (both of which are synonymous with the modern definition of aliens), as well as the ancient Earth, with its pre-Adamite society, is accurate, then it can safely be postulated that at least some angels faithful to God utilize advanced technology. They probably don't have to have it in many cases, but they use it in order to supplement or enhance their abilities. Consider the angel who guarded the entrance to the Garden of Eden and the tree of life after Adam and Eve were kicked out of the garden (Genesis 3:22–24). Did he absolutely have to have a flaming sword? What kind of weapon was that, anyway—a light saber? Why didn't the angel just wave his hand in order to prevent access to the tree of life? Could it be that he actually *needed* that sword?

> Genesis 3:22–24 (bold emphasis added)
>
> And the LORD God said, Behold, the man is become as one of us, to know good and evil: and now, lest he put forth his hand, and take also of the tree of life, and eat, and live for ever: Therefore the LORD God sent him forth from the Garden of Eden, to till the ground from whence he was taken. So He drove out the man; and **He placed at the east of the Garden of Eden Cherubims, and a flaming sword which turned every way**, to keep the way of the tree of life.

Could it be that the angels who were flying a sky chariot with fire all around it and who apprehended Elijah and took him to heaven, actually *needed* a vehicle capable of flying?

2 Kings 2:11–12 (bold emphasis added)

And it came to pass, as they still went on, and talked, that, behold, there appeared **a chariot of fire, and horses of fire, and parted them both asunder; and Elijah went up by a whirlwind into heaven.** And Elisha saw it, and He cried, My father, my father, the chariot of Israel, and the horsemen thereof.

The Bible speaks of angels walking (Genesis 19:1–3) and flying (Genesis 28:12; John 1:51; Revelation 14:6), but perhaps a vehicle gets certain angels where they want to go a little faster, just as an automobile gets people where they need to go a little faster, even though they don't absolutely have to have one, since they can always walk. Nevertheless isn't it true that technology is the natural expression of intelligence in the material universe? I acknowledge that these incredibly powerful, ultraintelligent beings could actually use something as antiquated as a chariot if they wanted to, but I think it's more likely it was another type of vehicle, and its description is given through the eyes of someone who knew of no other type of vehicle. That's why the power moving the chariot is referred to as a horse, albeit made of fire.

Now spirit horses in particular are actually mentioned in several places in scripture, and I see no reason why the inhabitants of the first heaven wouldn't ride them, especially when going into battle, just as we do in this world in certain places, though longer journeys probably require something a bit faster. Most likely the majority of the visions of spirit horses that are actually horses, and not symbolic imagery, are from horses in the first heaven right above our heads (2 Kings 6:17—could refer to both spacecraft and spirit horses, just as the armies of Earth may include soldiers mounted on horses and an assortment of vehicles). I doubt angels riding spirit horses cross the entire universe, though a spirit horse might have a few tricks up his sleeve. Overall, I think it would help people somewhat to get away from thinking of heaven and angels as purely magical and apply some logic to contemplating everything that scripture says about them.

The invisible realms can be known by that which is visible (Romans 1:20). Just as this world has a mix of the natural (creatures, technology, logic, principles, and so forth) and supernatural (powers, prophecy, miracles, and so forth), so it is in the heavens. A prime example of what I'm trying to explain can be derived by simply analyzing movies like *Star Wars* and *Star Trek* with a more eternal kingdom–oriented perspective. Scripture makes much more sense this way. These movies clearly depict the universe with a colorful mix of all manner of life-forms, technology, and beings with varying abilities, intelligence—and in some cases, supernatural power.

In *Star Wars*, remember that only the Jedi Knights imbued with the force (supernatural power) were able to make and use light sabers. Recall in *Star Trek* that most beings needed technology, while others, such as the Q, were so powerful they didn't need it at all. As such, I expect the heavens to be very diverse in these details, which are so clearly depicted in science fiction and in the Bible. The Bible supports a more science fiction type of perspective, rather than an antiquated fairytale paradigm. Reality at large is very diverse, yet interestingly, human reasoning has a natural tendency to learn about reality by accepting only one dimension of it at a time. People assimilate knowledge by categorizing and classifying, and in many cases, they unnecessarily eliminate even the possibility of that which is unknown.

With these details in mind, take note that no scripture ever mentions that angels are all-powerful; they have limits. Also realize that God doesn't do everything himself, nor is he a micromanager. God takes extensive risks in his delegations, and he constantly delegates to his angels many activities in the affairs of his creation. In light of these facts, it should be understood that the event of the rapture of the church referred to in numerous places in scripture will be a coordinated event with Jesus heading an extremely quick "flyby" with a massive fleet of angels within their sky chariots (1 Thessalonians 4:16–17; Psalms 68:17). It will be faithful angels who will be taking the physical bodies of saints from Earth, stealing them off to heaven, probably exactly the way they took Elijah (2 Kings 2:11), but on a much larger scale. What might this entail?

> 1 Thessalonians 4:16–17 (bold emphasis added)
>
> For the Lord himself shall descend from heaven with a shout, with the voice of the archangel, and with the trump of God: and **the dead in Christ shall rise first: Then we which are alive and remain shall be caught up together with them in the clouds, to meet the Lord in the air**: and so shall we ever be with the Lord.
>
> Psalms 68:17 (bold emphasis added)
>
> The **chariots of God are twenty thousand, even thousands of angels: the Lord is among them, as in Sinai, in the holy place.**

<u>2 Kings 2:11</u>

And it came to pass, as they still went on, and talked, that, behold, there appeared **a chariot of fire, and horses of fire, and parted them both asunder; and Elijah went up by a whirlwind into heaven**.

Consider for a moment the possibility of a fleet of intergalactic, interdimensional spaceships, capable of containing an excess of one billion people (all the saints of the past since the crucifixion, and all the saints presently on Earth). Imagine the sky proliferated with spacecraft, and hundreds of millions of people suddenly disappearing. The angels will arrive and quickly circle the globe in their sky chariots, taking millions of people as they pass by. I won't even bother to go into detail in describing Ezekiel's wheel (Ezekiel chapters 1 and 10), other than to say that there's every indication (at least to me) that technology seems to be in the picture.

During the rapture of the church, this massive arrival will most likely only be visible by the people being taken up. The purpose of this mission will not to be made manifest, but simply to gather the saints prior to the "great tribulation" (a time of unparalleled suffering, both in magnitude and intensity, unlike ever before in the history of the world). Note that while the rapture of the church may not be visible, the massive arrival to occur at the second coming of Christ will not be discreet at all. Isaiah 66:15–16 speaks of "chariots like a whirlwind" raining fire down on the Earth.

As for the rapture of the church, however—as suddenly as the angels arrive, they will disappear without a trace, in the twinkling of an eye (1 Corinthians 15:52), leaving the entire world to dither in complete dismay. Probably the only people who might see anything at all, if anyone, will be military personnel working at military installations like NORAD (North American Aerospace Defense Command). I doubt they'll say anything, but shortly after that time, their secrecy won't matter much anymore.

<u>Isaiah 66:1</u> (bold emphasis added)

For, behold, **the LORD will come with fire, and with his chariots like a whirlwind**, to render His anger with fury, and His rebuke with **flames of fire**. For **by fire and by His sword will the LORD plead with all flesh**: and the slain of the LORD shall be many.

1 Corinthians 15:52 (bold emphasis added)

In a moment, **in the twinkling of an eye**, at the last trump: for the trumpet shall sound, and **the dead shall be raised incorruptible, and we shall be changed**.

For a great many years, many Christians have argued over when and even if the rapture of the church will occur. While I hate to add my voice to a debate that will eventually solve itself (which is one of the reasons I sidestepped this topic in *Aliens in the Bible*), I have recently felt compelled to study this doctrine in depth, digging through various commentaries, especially those of Finis Dake's in *God's Plan for Man*. Another reason why I avoided the topic of the rapture of the church concerns one of the aspects of biblical interpretation that I don't like very much. Theology, in many respects, can be like politics; people can get extremely emotional about biblical interpretations if they believe the stakes are high enough. Unfortunately, all too often, ideas, beliefs, and concepts that aren't crucial to one's salvation are given too much importance, which results in a great deal of unwarranted bickering and emotionalism. I enjoy philosophizing, pondering, questioning, analyzing, and debating issues of faith, but only to the point that it's constructive, or at least entertaining. The fun quickly departs, however, when basic protocols of dignity and respect are breached.

Ironically, Christians sometimes stop acting like Christians when discussing their views of the Bible. Anyone familiar with theological critiques will see phrases like "grossly inaccurate," "false teachers," "false prophets," "promoting heresy," "deceivers," and the list goes on. Such comments often stem from legalism. If I have done this, please forgive me. I try to save such comments for stuff that really counts, like defending my belief that Jesus is the only way to the Father, and that God is not a created being (hence, alien). Even then, I try to focus more on pointing to the truth, rather than berating someone that needs to hear the truth without being ripped apart. This very issue of legalism is what upset Jesus the most concerning the behavior of the scribes and Pharisees. They would break all Ten Commandments in order to enforce one petty man-made rule.

If you, the reader, already have firm convictions regarding when or even if the rapture of the church will occur, before I lose your vote, I want to make myself perfectly clear concerning my mindset about this matter. I think the primary reason why the doctrine of the pretribulation rapture of the church is given so much weight by some people, is because it has been dubbed as an "escapist fantasy." Those who call the doctrine of the pretribulation rapture of the church a "heresy," among other colorful terms, do so because of their fear that the church is being setup for a fall. Christians are asleep, they say, and will not be prepared for the horrors about to engulf them, if they believe in the rapture of the church.

What do I have to say about this?

While the concern of a sleepy Christian church is legitimate, if the church isn't ready to sacrifice for Christ, then a problem exists regardless of the rapture of the church. Why waste time with arguing against the pretribulation rapture of the church, when the more direct issue has to do with fear and commitment?

Before conducting research into the topic of the rapture of the church, I was of the opinion that any perspective is suitable: if there is no rapture of the church, I will experience a powerful move of God in the church unlike any time in history, and if there is a rapture of the church, I will still experience God in an extremely intimate way. Any perspective (pretribulation, midtribulation, posttribulation), therefore, leaves incredible things to anticipate.

Now if I have a choice of whether to stay or go, I choose to stay. I aspire to have the same courage as the Christians in Hebrews 11:35 who sought a better resurrection.

> Hebrews 11:35 (bold emphasis added)
>
> Women received their dead raised to life again: **others were tortured, not accepting deliverance; that they might obtain a better resurrection**.

I'm inspired by these brave souls who sought a better resurrection. Perhaps I will never live up to this level of commitment, but it's certainly something that burns in my heart. Call it a bewildering pang in my soul, but I'd go through the same death that the first Christian martyr mentioned in the Bible went through, in order to experience that same glowing rapture that gave him the face of an angel when the heavens opened up to him (Acts 6:15, 7:54–60). It's much easier to say than to do, but if I weren't married with children, I'd be tempted to smuggle Bibles into China, because I want to have something to lie down at the feet of my savior when I see him. To date, I have lost very little for my Lord. He's more wonderful than any words can describe, and I yearn in earnest to present him with an offering that will bring tears to his eyes, because of the tears of joy that he's brought to my eyes.

> Acts 6:15
>
> And all that sat in the council, looking stedfastly on him, saw his face as it had been the face of an angel.

Acts 7:54–60

When they heard these things, they were cut to the heart, and they gnashed on him with their teeth. But he, being full of the Holy Ghost, looked up stedfastly into heaven, and saw the glory of God, and Jesus standing on the right hand of God, And said, Behold, I see the heavens opened, and the Son of man standing on the right hand of God. Then they cried out with a loud voice, and stopped their ears, and ran upon him with one accord, And cast him out of the city, and stoned him: and the witnesses laid down their clothes at a young man's feet, whose name was Saul. And they stoned Stephen, calling upon God, and saying, Lord Jesus, receive my spirit. And he kneeled down, and cried with a loud voice, Lord, lay not this sin to their charge. And when he had said this, he fell asleep.

I'm a member of VOM (Voice of the Martyrs), and nothing fires me up more than to read about the persecuted church at http://www.persecution.com.[1] I was turned on to VOM after reading an incredible book called *Jesus Freaks*. Another similar book I recommend is *Foxe's Book of Martyrs*. Christians are beaten, tortured, raped, sold into slavery, and murdered—EVERY single day, all around the world, for no other reason than being Christian. Close to 156,000 Christians were martyred for their faith in 1998.[2] This number is expected to grow with each passing year.

VOM currently operates in forty nations around the world, where such atrocities are common. China is currently the number one human rights violator in the world,[3] which is why I would deliver Bibles there. I heard a missionary say on a video I watched in my church the other night that one of the prerequisites for being a home-church pastor in China is serving time in jail. Now that's an interesting job resume! In America, we put "Graduated with honors at theological seminary bla bla bla," while in China, they put "Look forward to more time in jail."

In Pakistan, Christians will routinely have an arm (or some other appendage) chopped off as one of the various tortures to choose from.[4] The faithful step forward, losing their arms, and give praise to God because they have one arm left in which to give him praise!

There's nothing wrong with fear and commitment in the Pakistan church, and if a Christian in Pakistan voiced a belief in the rapture of the church, who would object? What theological concept is worth arguing about with a Christian who lost an arm because of his or her faith? This is very different from the theological bickering I alluded to previously. For anyone who wants to get into an emotional, theological debate about anything, I highly recommend spending some time in the mission field of a country like Pakistan, because actions speak much louder than words.

What makes Christians so brave like those in China and Pakistan? The love of Christ is a source of infinite power. The fearlessness exhibited in these Christians has nothing to do with someone withholding the doctrine of the rapture from their ears. They're on fire for God, and nothing will extinguish them. When I hear about these saints, it makes me want to be a martyr too.

When I was young, I used to be terrified at the thought of the Antichrist. Now, however, I'd relish the opportunity to boldly, publicly deny him to his face, and then sing glory to God while being killed for doing so. This attitude is nothing I want to *impose* on anyone; it's simply the kind of Christian character to wish to have.

So there's my disclaimer about promoting an escapist doctrine. My mindset will always be in favor of sacrificing for Jesus, and my belief in the pretribulation rapture of the church has nothing to do with wanting to escape persecution. I might not be volunteering for persecution at this time in my life, but if it comes my way, I hope I won't cower away from it. I also advise anyone who believes in the rapture of the church because of a fear of persecution to watch *The Passion of the Christ*, and to read books like *Jesus Freaks*. Jesus died not only to atone for all the sin in the world, but also to show us how to die as well. Taking all of this information into account, I will now discuss why I believe there will be a pretribulation rapture of the church, because that's what I see in scripture, and not because I don't want to endure persecution.

In the process of researching the rapture of the church, I have found a few reasons why there is so much confusion about this future event. Many scriptures referring to the rapture of the church are very clear, however, and leave little room for misinterpretation. I am now of the opinion that the rapture of the church will definitely happen, and I believe it will occur before the great tribulation (which unbelievably, will actually be worse than the tribulation currently going on in the world). If it doesn't, that's fine, because all Christians should be willing to suffer for Christ, following his example and being pleased to have something extremely tangible to offer him. I think scripture indicates that those who will be suffering in the great tribulation, however, are those who aren't Christians now, but will become Christians after the rapture of the church.

For those who have no interest in knowing when the rapture of the church will occur, I recommend simply skipping to the next chapter. But for those who want to know more about this "rapture of the church thing"—read further!

1. The Pretribulation Rapture of the Church Is Consistent with the Heart of God

Christian terms, such as "pretrib," "midtrib," and "post-trib," are spoken of in Christian circles, referring to the rapture of the church as happening either before the great tribulation, during the middle of the great tribulation, or at the end of the great tribulation. After studying all of these views, I have found the pretribulation view to be perfectly consistent with the heart of God, as well as with scripture, while all the other views are contradictory to both the heart of God and scripture.

The simplest argument of all concerning the timing of the rapture of the church comes from knowing the heart of God. The Old Testament gives what I believe to be an excellent example of a foreshadowing of the rapture of the church. In Genesis 18:23–25, before God destroyed Sodom and Gomorrah, Abraham asked God a bold question: Is it just that the righteous be destroyed with the unrighteous? God answered Abraham's question by taking Lot and his family out of Sodom and Gomorrah in the nick of time (Genesis 19:1–29). This is God; it's who he is.

> Genesis 18:22–25 (bold emphasis added)
>
> And the men turned their faces from thence, and went toward Sodom: but Abraham stood yet before the LORD. And Abraham drew near, and said, **Wilt thou also destroy the righteous with the wicked?** Peradventure there be fifty righteous within the city: wilt thou also destroy and not spare the place for the fifty righteous that are therein? **That be far from thee to do after this manner, to slay the righteous with the wicked**: and that the righteous should be as the wicked, that be far from thee: **Shall not the Judge of all the earth do right?**

Now it's true that God allows suffering and severe hardships to befall his children, and even his own Son for that matter; there's no doubt about that, but the suffering poured out is not coming from God himself. It's coming from Satan, this sinful world, and the consequences of our own sins. Furthermore God never appoints anyone to meaningless suffering. His love and glory are always at the center of all of his actions. When people are martyred for Christ, the world sees the suffering that Christians are willing to go through because of their love for God. Concerning wrath issued directly from God himself, this is not the case. How could God receive glory by pouring out his own wrath on his own obedient children? A child beaten up at school because he's defending someone or standing up for his family's honor brings glory to the father of the family, but the father beating all of his children to death because of the disobedience of *some* of them does what? This is commonsense stuff, and I don't need any commentary to understand it.

The only time suffering is in the will of God is when it benefits other people or gives glory to God in some way (as implied in Hebrews 11:35–40).

> Hebrews 11:35–40 (bold emphasis added)
>
> Women received their dead raised to life again: **and others were tortured, not accepting deliverance; that they might obtain a better resurrection**: And others had trial of cruel mocking and scourging, yea, moreover of bonds and imprisonment: They were stoned, they were sawn asunder, were tempted, were slain with the sword: they wandered about in sheepskins and goatskins; being destitute, afflicted, tormented; (Of whom the world was not worthy:) they wandered in deserts, and in mountains, and in dens and caves of the earth. And these all, having obtained a good report through faith, received not the promise: God having provided some better thing for us, that they without us should not be made perfect.

When it comes to suffering, God leads by example in an extreme way. Jesus's death glorified God the Father, because he took upon himself the punishment we all deserve in order to save us from hell (John 3:16). God wants to save us, not kill us! He's not willing that any should perish (2 Peter 3:9).

> John 3:14–19 (bold emphasis added)
>
> And as Moses lifted up the serpent in the wilderness, even so must the Son of man be lifted up: That whosoever believeth in him should not perish, but have eternal life. For **God so loved the world, that He gave his only begotten Son, that whosoever believeth in him should not perish, but have everlasting life. For God sent not his Son into the world to condemn the world; but that the world through him might be saved**. He that believeth on him is not condemned: but he that believeth not is condemned already, because he hath not believed in the name of the only begotten Son of God. And this is the condemnation, that light is come into the world, and men loved darkness rather than light, because their deeds were evil.

> 2 Peter 3:9 (bold emphasis added)
>
> **The Lord is** not slack concerning his promise, as some men count slackness; but is **longsuffering to us-ward, not willing that any should perish,** but that all should come to repentance.

As for the martyrs who have followed Jesus's example even to this day, they too suffer for Christ's sake, because they love him too much to deny him, or even hide their faith in him. As the apostle Paul stated, the weight in glory received from God is far more than the price paid in suffering (2 Corinthians 4:17).

2 Corinthians 4:17

> For our light affliction, which is but for a moment, works for us a far more exceeding and eternal weight of glory...

God receives glory in martyrdom and sacrifice for the good of all, but he does not receive glory from personally killing his own faithful children. Going back to the destruction of Sodom and Gomorrah, if Lot and his family died in that destruction, God would not have received any glory from it. With that same commonsense reasoning, doesn't it make sense that God would save his faithful children from the great tribulation in the near future?

Now there will be people who will suffer for Christ during the great tribulation, but this is because of their own lack of faith preceding the rapture of the church. This is a common misunderstanding some people have. Some advocates of the rapture have a fire-and-brimstone message concerning the rapture of the church, stating that all will be hopeless for those who are left behind, and that it will be too late for them to accept Christ after the rapture of the church. This is unscriptural, because many will be saved during the great tribulation (this is actually one of the primary reasons God is bringing his judgment to Earth—it's a last-chance wake-up call for the sake of mercy), though their faith will be tried to utter extremes.

The tribulation saints, as they are called, wouldn't have to deal with the great tribulation if they were "in Christ" before the rapture of the church (2 Corinthians 5:17; 1 Thessalonians 4:16).

2 Corinthians 5:17 (bold emphasis added)

> Therefore if any man be **in Christ**, he is a new creature: old things are passed away; behold all things are become new.

1 Thessalonians 4:16 (bold emphasis added)

> For the Lord Himself shall descend from heaven with a shout, with the voice of the archangel, and with the trump of God: and the dead **in Christ** shall rise first...

In essence they will be forced to endure some of the hardships of that time, because of their unbelief prior to the rapture of the church. The brunt of their suffering will come especially from the Antichrist, but God in his infinite mercy will still allow them to accept him even after the rapture of the church (tribulation saints are saved—Revelation 6:9–11, 13:7–18, 15:2–4, 20:4–6; and the 144,000 Messianic Jews will all be saved—Revelation 7:4, 14:1–3).

God will also withhold his wrath from those who are saved during the great tribulation, though he won't restrain the Antichrist. We know this from the fact that all of the tribulation saints mentioned in scripture are those who will be slain by the Antichrist (Revelation 13:7–18; 15:2–4; 20:4–6), rather than the many terrible punishments issued directly from God. And even though it will be their initial lack of faith that results in their suffering in the tribulation, God will even bestow upon them a special blessing because of their dedication to him during this tumultuous time (Revelation 14:13).

> Revelation 14:13
>
> And I heard a voice from heaven saying unto me, Write, Blessed are the dead who die in the Lord from henceforth: Yea, says the Spirit that they may rest from their labors; and their works do follow them.

Concerning the 144,000 Jews mentioned in the book of Revelation, who will be the Jewish saints (Messianic Jews) during the time of the great tribulation, the first four trumpet angels will be ordered not to hurt Earth, trees, grass, and seas until the 144,000 are sealed. This seal will serve as a protective barrier for the Jews to protect them from the trumpet judgments that follow. Special direction will then be given to the fifth and sixth trumpet angels to hurt only those men who won't be sealed by God's faithful angels (Revelation 9:4). After this, the 144,000 Jews will again be protected, this time from demon creatures loosed in Revelation 9:1–12, and the 200 million demon horsemen liberated out of the same Abyss to slay one-third of all men at a certain hour.

> Revelation 9:4
>
> And it was commanded them that they should not hurt the grass of the earth, neither any green thing, neither any tree; but only those men which have not the seal of God in their foreheads.

With all these facts taken together, it should be easy to conclude that the wrath of God during the great tribulation is poured out upon the world *because* of the unjust killing of his saints, not because he just wants to hurt everyone in the entire world, *including* his saints. It should be well understood that God has not appointed his own children to wrath (Ephesians 2:1–9). Jesus even told his disciples that everyone should strive to be worthy of escaping the tribulation (Luke 21:34–36). Why would he say this if there were no hope of escape? To escape the tribulation means to get out before it happens—which is common sense, in my opinion, but some people need more explanation. That is why my argument about the rapture of the church occurring before the great tribulation has logical reasoning with scriptural support behind it as well, which I obtained from Finis Dake's *God's Plan for Man*, and I will now divulge.

> Ephesians 2:1–9 (bold emphasis added)
>
> And you hath He quickened, who were dead in trespasses and sins; Wherein in time past ye walked according to the course of this world, according to the prince of the power of the air, the spirit that now works in the children of disobedience: Among whom also we all had our conversation in times past in the lusts of our flesh, fulfilling the desires of the flesh and of the mind; and **were by nature the children of wrath, even as others. But God, who is rich in mercy, for his great love wherewith He loved us, Even when we were dead in sins, hath quickened us together with Christ, (by grace ye are saved;) And hath raised us up together, and made us sit together in heavenly places in Christ Jesus**: That in the ages to come He might show the exceeding riches of his grace in his kindness toward us through Christ Jesus. For by grace are ye saved through faith; and that not of yourselves: it is the gift of God: Not of works, lest any man should boast.

2. The Pretribulation Rapture of the Church Is Consistent with Scripture

The main mistake about the rapture of the church doctrine that confuses people more often than anything else is the mixing up of the rapture of the church with the second coming of Christ. The rapture of the church and the second coming of Christ are two separate events that have a few similar elements. Despite their similarities, they are definitely two distinct events that fall in line with specific periods.

The second element of confusion concerning the rapture of the church has to do with its definition. The reason I've been saying the phrase "rapture of the church" is because there will be more than one rapture. This alone can bring confusion, but it shouldn't, in light of the fact that each rapture is for a specific group of people (specified with the term "each in his own order"—1 Corinthians

15:20–23). The rapture that people argue the most about is the one specifically for the church, because this one will cause the biggest shock to the sleepy, modern world, but it is frequently mixed up with the other raptures that follow it, especially that of the tribulation saints. Now I will provide the details and clarification about these two common misconceptions of the rapture.

> 1 Corinthians 15:20–23 (bold emphasis added)
>
> But now is Christ rose from the dead, and become the first fruits of them that slept. For since by man came death, by man came also the resurrection of the dead. For as in Adam all die, even so **in Christ shall all be made alive**. But **every man in his own order: Christ the first fruits; afterward they that are Christ's at his coming**.

2.1. Mixing Up the Second Advent of Christ with the Rapture of the Church

The rapture of the church is a distinct event in itself and takes place at least seven years before the second coming of Christ (Daniel 9:27). The rapture takes place before the tribulation (which will be explained in detail below), and the second coming after the tribulation. The rapture of the church is the time when Christ comes for the saints that are physically alive, and for the physical bodies of those who have been dead in Christ (1 Thessalonians 4:13–17). The second coming, on the other hand, is when he comes back to Earth with all of the resurrected saints in order to drop them off (Zechariah 14:1–5; Jude 14; Revelation 19:11–21). At the rapture, Christ takes all of the resurrected saints to heaven (1 Thessalonians 3:13, 4:16; Colossians 3:4), and at the second coming, he leaves heaven with them (Revelation 19:1–21). At the rapture of the church, Christ does not come down to Earth, but flies in the sky (1 Thessalonians 4:16), while at the second coming, he does come down to Earth (Zechariah 14:4; Matthew 24:29–31).

> Daniel 9:27*
>
> And he shall confirm the covenant with many for one week: and in the midst of the week he shall cause the sacrifice and the oblation to cease, and for the overspreading of abominations he shall make it desolate, even until the consummation, and that determined shall be poured upon the desolate.
>
> *Each day in Daniel's seventieth week represents one year. Prior to this seven year period is the rapture of the church, and at the conclusion of this last week in Daniel's vision, Jesus returns to the Earth in the second advent.*

1 Thessalonians 4:13–17—rapture of the church* (bold emphasis and bracketed comments added)

But I would not have you to be ignorant, brethren, concerning them which are asleep, that ye sorrow not, even as others which have no hope. For if we believe that Jesus died and rose again, even so **them also which sleep in Jesus** will God bring with him [Jesus will bring the spirits of Christians from ages past, in order to reunify them with their resurrected, physical bodies]. For this we say unto you by the word of the Lord, that we which are alive and remain unto the coming of the Lord shall not prevent them which are asleep. For the Lord himself **shall descend from heaven with a shout**, with the voice of the archangel, and with the trump of God: and **the dead in Christ shall rise first: Then we which are alive and remain shall be caught up together with them in the clouds, to meet the Lord in the air**: and so shall we ever be with the Lord.

It would make no sense to pull everyone up into the air, and then put everyone back down on the Earth again. Some have postulated that this is a customary Jewish greeting similar to the greeting of a famous dignitary, but there is no hard evidence to support such an interpretation. Would God really levitate several hundred million people into the air in a global, public display, just to say "Hi," and then set them back down on the Earth again? I think this scripture makes more sense when understood as being the first part of the same event spoken of in 1 Thessalonians 3:13, mentioned below.

1 Thessalonians 3:13—rapture of the church* (bold emphasis added)

To the end He may establish your hearts unblameable in holiness **before God, even our Father, at the coming of our Lord Jesus Christ with all his saints**.

In this scripture, Jesus comes not to the Earth, but from the Earth, going to heaven, before God the Father, with all of his saints, in order to present them to God the Father.

Zechariah 14:1–5—second advent (bold emphasis added)

Behold, the day of the LORD cometh, and thy spoil shall be divided in the midst of thee. For I will gather all nations against Jerusalem to battle; and the city shall be taken, and the houses rifled, and the women ravished; and half of the city shall go forth into captivity, and the residue of the people shall not be cut off from the city. Then shall the LORD go forth, and fight against those nations, as when He fought in the day of battle. And his feet shall stand in that day upon the mount of Olives, which is before Jerusalem on the east, and the mount of Olives shall cleave in the midst thereof toward the east and toward the west, and there shall be a very great valley; and half of the mountain shall remove toward the north, and half of it toward the south. And ye shall flee to the valley of the mountains; for the valley of the mountains shall reach unto Azal: yea, ye shall flee, like as ye fled from before the earthquake in the days of Uzziah king of Judah: **and the LORD my God shall come, and all the saints with thee**.

Jude 1:14–15—second advent (bold emphasis added)

And Enoch also, the seventh from Adam, prophesied of these, saying, **Behold, the Lord cometh with ten thousands of his saints**, To execute judgment upon all, and to convince all that are ungodly among them of all their ungodly deeds which they have ungodly committed, and of all their hard speeches which ungodly sinners have spoken against him.

Revelation 19:11–21—second advent (bold emphasis added)

And I saw heaven opened, and behold a white horse; and He that sat upon him was called Faithful and True, and in righteousness He doth judge and make war. His eyes were as a flame of fire, and on his head were many crowns; and He had a name written, that no man knew, but Him. And He was clothed with a vesture dipped in blood: and his name is called The Word of God. **And the armies which were in heaven followed him upon white horses, clothed in fine linen, white and clean**. And out of his mouth goes a sharp sword, that with it He should smite the nations: and He shall rule them with a rod of iron: and He treads the winepress of the fierceness and wrath of Almighty God. And He hath on his vesture and on his thigh a name written, KING OF KINGS, AND LORD OF LORDS. And I saw an angel standing in the sun; and He cried with a loud voice, saying to all the fowls that fly in the midst of heaven, Come and gather yourselves together unto the supper of the great God; That ye may eat the flesh of kings, and the flesh of captains, and the flesh of mighty men, and the flesh of horses, and of them that sit on them, and the flesh of all men, both free and bond, both small and great. And I saw the beast, and the kings of the earth, and their armies, gathered together to make war against him that sat on the horse, and against his army. And the beast was taken, and with him the false prophet that wrought miracles before him, with which he deceived them that had received the mark of the beast, and them that worshipped his image. These both were cast alive into a lake of fire burning with brimstone. And the remnant were slain with the sword of him that sat upon the horse, which sword proceeded out of his mouth: and all the fowls were filled with their flesh.

Colossians 3:4—rapture of the church (bold emphasis added)

When Christ, who is our life, shall appear, **then shall ye also appear with him in glory**.

Matthew 24:29–31—second advent

> Immediately after the tribulation of those days shall the sun be darkened, and the moon shall not give her light, and the stars shall fall from heaven, and the powers of the heavens shall be shaken: And then shall appear the sign of the Son of man in heaven: and then shall all the tribes of the earth mourn, and they shall see the Son of man coming in the clouds of heaven with power and great glory. And He shall send his angels with a great sound of a trumpet, and they shall gather together his elect from the four winds, from one end of heaven to the other.

Unfortunately many people jumble all of these scriptures together to support one conclusion or another—but missing only one or two scriptures can end up resulting in a contradiction. It helps to be familiar with the entire Bible, and to be capable of comparing and contrasting all these scriptures with logical consistency, in order to see that the second coming of Christ and the rapture of the church are in fact two different events. Following is a detailed, brief analysis of the rapture of the church, and then the second advent of Christ coming to Earth, after which it should be clear that these are two distinctly different events.

2.1.1. Details Concerning the Rapture of the Church

To begin with, the rapture of the church is defined as the catching up of all true believers of Christ to meet the Lord in the air. Clearly this event is prophesied to occur in the future in 1 Thessalonians 4:13–17; 1 Corinthians 15:23, 51–58; Philippians 3:20–21; John 14:1–3; Luke 21:34–36; Colossians 3:4, and others.

> 1 Thessalonians 4:13–17 (bold emphasis added)
>
> But I would not have you to be ignorant, brethren, concerning them which are asleep, that ye sorrow not, even as others which have no hope. For if we believe that Jesus died and rose again, even so **them also which sleep in Jesus will God bring with him**. For this we say unto you by the word of the Lord, that **we which are alive and remain unto the coming of the Lord shall not prevent them which are asleep. For the Lord himself shall descend from heaven with a shout, with the voice of the archangel, and with the trump of God: and the dead in Christ shall rise first: Then we which are alive and remain shall be caught up together with them in the clouds, to meet the Lord in the air: and so shall we ever be with the Lord.**

1 Corinthians 15:23

But every man in his own order: Christ the first fruits; afterward they that are Christ's at his coming.

1 Corinthians 15:51–58 (bold emphasis and bracketed comments added)

Behold, I show you a mystery; **we shall not all sleep** [many will never physically die], **but we shall all be changed**, in a moment, in the twinkling of an eye, at the last trump: for the trumpet shall sound, and **the dead shall be raised incorruptible**, and **we shall be changed**. For this corruptible must put on incorruption, and this mortal must put on immortality. So when this corruptible shall have put on incorruption, and this mortal shall have put on immortality, and then shall be brought to pass the saying that is written, Death is swallowed up in victory. O death, where is thy sting? O grave, where is thy victory? The sting of death is sin; and the strength of sin is the law. But thanks be to God, which gives us the victory through our Lord Jesus Christ. Therefore, my beloved brethren, be ye steadfast, unmovable, always abounding in the work of the Lord, forasmuch as ye know that your labor is not in vain in the Lord.

Philippians 3:20–21 (bold emphasis added)

For our conversation is in heaven; from whence also we look for the Savior, the Lord Jesus Christ: **Who shall change our vile body, that it may be fashioned like unto his glorious body**, according to the working whereby He is able even to subdue all things unto himself.

John 14:1–3 (bold emphasis added)

Let not your heart be troubled: ye believe in God, believe also in me. **In my Father's house are many mansions**: if it were not so, I would have told you. **I go to prepare a place for you**. And if I go and prepare a place for you, **I will come again, and receive you unto myself; that where I am, there ye may be also**.

Luke 21:34–36 (bold emphasis added)

And take heed to yourselves, lest at any time your hearts be overcharged with surfeiting, and drunkenness, and cares of this life, and so that day come upon you unawares. For as a snare shall it come on all them that dwell on the face of the whole earth. **Watch ye therefore, and pray always, that ye may be accounted worthy to escape all these things that shall come to pass**, and to stand before the Son of man.

Colossians 3:4 (bold emphasis added)

When **Christ, who is our life, shall appear, then shall ye also appear with him in glory**...

For those who can read these scriptures, especially 1 Thessalonians 4:13–17, and 1 Corinthians 15:51–58, and not believe that the Bible clearly speaks of a rapture, I'm at a loss about what to say. It's too clearly stated to be confusing. Fortunately, I've found that most Christians believe that a rapture will occur; it's just the time frame of the rapture that is usually in contention.

Following are the distinguishing characteristics of the rapture of the church:

- The rapture of the church is a New Testament doctrine never revealed to anyone in the Old Testament; it was first alluded to by Jesus (Luke 21:34–36—see above), and then revealed to Paul in detail as a special revelation (1 Corinthians 15:51–58—see above; 2 Thessalonians 2:6–8).

  2 Thessalonians 2:6–8 (bold emphasis and bracketed comments added)

  And now ye know what withholds that he might be revealed in his time. For the mystery of iniquity doth already work: only he who now letteth will let, **until he [the church] be taken out of the way [raptured]**. And then shall that Wicked [the Antichrist] be revealed, whom the Lord shall consume with the spirit of his mouth, and shall destroy with the brightness of his coming...

- The rapture of the church can occur at any time and requires no fulfillment of prophecy or any sign (early Christians were looking for the rapture of the church at any time in their day without any particular prophecy being fulfilled—Philippians 3:20–21; Luke 21:34–36—see above; Titus 2:11–13).

  Titus 2:11–13 (bold emphasis added)

  For the grace of God that brings salvation hath appeared to all men, Teaching us that, denying ungodliness and worldly lusts, **we should live soberly, righteously, and godly, in this present world; Looking for that blessed hope, and the glorious appearing of the great God and our Savior Jesus Christ**...

- The rapture of the church will occur before the great tribulation (Luke 21:34–36; 2 Thessalonians 2:6–8; Revelation 1:19; 4:1—further explanation of this conclusion is given later in this chapter).

- Christ will *appear* visibly only to the saints in the air during the rapture of the church. The Greek word for "appear" is "phaneros," means "to shine," "be apparent," "manifest," or "be seen," found in John 2:28, 3:2; 1 Peter 5:4; Colossians 3:4. Christ is to appear to the saints in the air at the rapture, but nothing is said of his appearing to the rest of Earth.

- During the rapture of the church, Christ will never set foot on Earth, but rather fly above it in the air. As soon as all the dead in Christ have met him in the air, he will return to heaven with them to present them blameless before God (John 14:1–3; 1 Thessalonians 3:13, 4:16–17, and others).

John 14:1–3 (bold emphasis added)

Let not your heart be troubled: ye believe in God, believe also in me. **In my Father's house are many mansions**: if it were not so, I would have told you. **I go to prepare a place for you.** And if I go and prepare a place for you, **I will come again, and receive you unto myself; that where I am, there ye may be also.**

1 Thessalonians 3:13 (bold emphasis added)

To the end He may establish your hearts **unblameable in holiness before God, even our Father, at the coming of our Lord Jesus Christ with all his saints**.

2.1.2. Details Concerning the Second Advent of Christ

The second advent of Christ is defined as the physical return of Christ in bodily form to Earth (Zechariah 14:1–9; Matthew 24:29–31; Acts 1:11; Revelation 19:11–21) to establish his kingdom where he will rule on Earth over all nations forever (Daniel 2:44–45, 7:13–14, 7:18; Isaiah 9:6–7; Luke 1:32–33; Revelation 11:15).

Matthew 24:29–31 (bold emphasis added)

Immediately after the tribulation of those days shall the sun be darkened, and the moon shall not give her light, and the stars shall fall from heaven, and the powers of the heavens shall be shaken: **And then shall appear the sign of the Son of man in heaven: and then shall all the tribes of the earth mourn, and they shall see the Son of man coming in the clouds of heaven with power and great glory**. And He shall send his angels with a great sound of a trumpet, and they shall gather together his elect from the four winds, from one end of heaven to the other.

Acts 1:11

Which also said, Ye men of Galilee, why stand ye gazing up into heaven? This same Jesus, who is taken up from you into heaven, shall so come in like manner as ye have seen him go into heaven.

Revelation 19:11–21 (bold emphasis added)

And I saw heaven opened, and behold a white horse; and He that sat upon him was called Faithful and True, and in righteousness He doth judge and make war. His eyes were as a flame of fire, and on his head were many crowns; and He had a name written, that no man knew, but Him. And He was clothed with a vesture dipped in blood: and his name is called The Word of God. **And the armies which were in heaven followed him upon white horses**, clothed in fine linen, white and clean. And out of his mouth goes a sharp sword, that with it He should smite the nations: and He shall rule them with a rod of iron: and He treads the winepress of the fierceness and wrath of Almighty God. And He hath on his vesture and on his thigh a name written, KING OF KINGS, AND LORD OF LORDS. And I saw an angel standing in the sun; and He cried with a loud voice, saying to all the fowls that fly in the midst of heaven, Come and gather yourselves together unto the supper of the great God; That ye may eat the flesh of kings, and the flesh of captains, and the flesh of mighty men, and the flesh of horses, and of them that sit on them, and the flesh of all men, both free and bond, both small and great. **And I saw the beast, and the kings of the earth, and their armies, gathered together to make war against him that sat on the horse, and against his army.** And the beast was taken, and with him the false prophet that wrought miracles before him, with which he deceived them that had received the mark of the beast, and them that worshipped his image. These both were cast alive into a lake of fire burning with brimstone. And the remnant were slain with the sword of him that sat upon the horse, which sword proceeded out of his mouth: and all the fowls were filled with their flesh.

Daniel 2:44–45 (bold emphasis added)

And in the days of these kings shall **the God of heaven set up a kingdom, which shall never be destroyed: and the kingdom shall not be left to other people, but it shall break in pieces and consume all these kingdoms, and it shall stand for ever**. Forasmuch as thou sawest that the stone was cut out of the mountain without hands, and that it brake in pieces the iron, the brass, the clay, the silver, and the gold; the great God hath made known to the king what shall come to pass hereafter: and the dream is certain, and the interpretation thereof sure.

Isaiah 9:6–7 (bold emphasis added)

For unto us a child is born, unto us a son is given: and the government shall be upon his shoulder: and his name shall be called Wonderful, Counselor, The mighty God, The everlasting Father, The Prince of Peace. Of the increase of his government and peace there shall be no end, upon the throne of David, and upon his kingdom, to order it, and to establish it with judgment and with justice from henceforth even for ever. The zeal of the LORD of hosts will perform this.

Luke 1:32–33 (bold emphasis added)

He shall be great, and shall be called the Son of the Highest: and the Lord God shall give unto him the throne of his father David: And **He shall reign over the house of Jacob for ever; and of his kingdom there shall be no end.**

When he returns in the second advent, he will return during the battle of Armageddon, which will be waged between the armies of heaven under him, and the armies of Earth under Antichrist (Zechariah 14; Joel 3; Revelation 19:11–21).

Joel 3:15–16

The sun and the moon shall be darkened, and the stars shall withdraw their shining. The LORD also shall roar out of Zion, and utter his voice from Jerusalem; and the heavens and the earth shall shake: but the LORD will be the hope of his people, and the strength of the children of Israel.

On this occasion, Jesus's return will be visible to all of humanity (Hebrews 9:28, Isaiah 66:15–16, and other passages where it is obvious that God's activities won't be hidden from the people of Earth). And just as it is so popularly predicted in Hollywood, these otherworldly beings in their magnificent assembly of unbelievably awesome arsenal will rain down complete and utter destruction on the inhabitants of Earth below (Isaiah 66:15–16)—except, unlike in Hollywood movies, most of the people of Earth will be the bad guys in this case!

Isaiah 66:15–16 (bold emphasis added)

For, behold, **the LORD will come with fire, and with his chariots like a whirlwind**, to render his anger with fury, and his rebuke with flames of fire. For by fire and by his sword will the LORD plead with all flesh: and **the slain of the LORD shall be many.**

<u>Hebrews 9:28</u> (bold emphasis added)

So Christ was once offered to bear the sins of many; and **unto them that look for him shall He appear the second time** without sin unto salvation.

Interestingly, movies such as *Independence Day*, *Signs*, and *War of the Worlds*, becoming increasingly popular these days, are prophetic to a certain extent, yet frequently depict otherworldly invaders as the bad guys, which will not always be the case when Jesus returns to Earth to establish his kingdom, as shown in Isaiah 66:15–16. (Fallen angels will have their place in the skies as well, which I will discuss in the next chapter, but most of the world will actually perceive everything backward, claiming allegiance with the Antichrist instead of with God.)

The Antichrist will be very successful in convincing the people of Earth to unite together to fight this massive invasion from God, for he (the Antichrist) will convince everyone that this invasion is from a terrible enemy intent on global destruction. Exactly the opposite will be true, however. God's invasion will be defensive rather than offensive. His primary objective will be to defend Israel from the forces of the Antichrist, and that he will do quite successfully. The battle of Armageddon will be a bloody mess, that's for sure!

The movies I just mentioned are no accident. As I already stated, I believe it is true that the science fiction authors of today are right about how the heavens will be shaken (Matthew 24:29); in this respect, many Hollywood movies are prophetic. The skies may very well be proliferated with otherworldly spacecraft during the days of the Antichrist, just as they most likely were during the days of Noah (Genesis 6; Matthew 24:37; Luke 17:26). The greatest threat of all on Earth in those days, however, will be that of the deception coming from the Antichrist, and the fallen angels and demonic beings aligned with him, rather than physical destruction. Physical destruction only destroys the body *temporarily*, but deception creates a condition where a person's soul and spirit are destroyed for all eternity (Matthew 10:28).

<u>Matthew 24:29</u>

Immediately after the tribulation of those days shall the sun be darkened, and the moon shall not give her light, and the stars shall fall from heaven, and the powers of the heavens shall be shaken…

<u>Genesis 6:1–6</u>*

And it came to pass, when men began to multiply on the face of the earth, and daughters were born unto them, that the sons of God saw the daughters of men that they were fair; and they took them wives of all which they chose. And the LORD said, My spirit shall not always strive with man, for that he also is flesh: yet his days shall be an hundred and twenty years. There were giants in the earth in those days; and also after that, when the sons of God came in unto the daughters of men, and they bare children to them, the same became mighty men which were of old, men of renown. And GOD saw that the wickedness of man was great in the earth, and that every imagination of the thoughts of his heart was only evil continually. And it repented the LORD that He had made man on the earth, and it grieved him at his heart.

Fallen angels openly cohabitating with humans and interbreeding with them is the biggest distinguishing factor that differentiates the days of Noah from today. These unnatural unions were the primary impetus to the profound corruption of the entire world in those days.

Matthew 24:37

But as the days of Noah were, so shall also the coming of the Son of man be.

Luke 17:26

And as it was in the days of Noah, so shall it be also in the days of the Son of man.

Matthew 10:28 (bold emphasis added)

And **fear not them which kill the body, but are not able to kill the soul: but rather fear him which is able to destroy both soul and body in hell.**

Just as the rapture has its distinguishing characteristics, so does the second coming, or advent, of Christ. Following is a brief list of these characteristics:

- The second advent of Christ is an Old Testament doctrine given by Enoch (Genesis 5:21–24), Jacob (Genesis 49:10), Balaam (Numbers 24:7, 17–19), Isaiah (Isaiah 59:20, 63:1–5, 66:14–16), Jeremiah (Jeremiah 23:5–6, 25:30–33), Ezekiel (Ezekiel 34:23–29, 37:17–29, 43:7), Daniel (Daniel 2:44–45, 7:13–14), Hosea (Hosea 2:18–23, 3:4–5), Joel (Joel 2:28–3:21), Amos (Amos 5:15–21), Micah (Micah 1:3–4, 2:12–13, 4:1–5:7), Nahum (Nahum 1:5–6), Habakkuk (Habakkuk 2:13–14), Zephaniah (Zephaniah 1:14–18, 3:8–9), Haggai (Haggai 2:6–7, 21–23), Zechariah (Zechariah 2:10–13, 3:8, 6:12–13, 8:3–23, 12:4–14, 13:1–9, 14:1–21), and Malachi (Malachi 3:1–4:6), as well as a New Testament doctrine given by Jesus (Matthew 16:27, 24:1–25:46, Luke 17:22–37, 21:1–33), Peter (Acts 3:21; 2 Peter 1:16; 3:3–9), Paul (Romans 11:26–27; 2 Thessalonians 1:7–10, 2:1–8; Hebrews 9:28), Jude (Jude 14–15), and John (Revelation 1:7, 19:11–21). Not one of these scriptures refers to the rapture of the church. Because there are so many scriptures listed here, I'm not going to reference them as I've been doing throughout this book, because they would take up too much space. I recommend either downloading e-Sword or maybe even doing something crazy like investing a dollar to purchase a Bible at the local dollar store, where Bibles are usually sold for a buck.

- While the rapture of the church can occur at any time, *many* things *must* happen before the second coming of Christ. This is one of the primary reasons the second coming of Christ cannot be the rapture of the church. Many signs either have already occurred or are currently in progress. Some of these include conflicts with people, such as wars and rumors of wars (Matthew 24:6; Mark 13:7; Luke 21:9), nations against nations (Matthew 24:7; Mark 13:8, and many others), and the persecution of the Jews by all nations (Matthew 24:9; Mark 13:9–11; Luke 21:12; Joel 3; Zechariah 14; Revelation 12:13–17). Other signs include famines (Matthew 24:7; Mark 13:8; Luke 21:11; Revelation 6:7–8), pestilences (Matthew 24:7; Mark 13:8; Luke 21:11; Revelation 6:12–17, and many others), earthquakes (Matthew 24:7; Mark 13:8, and many others), an abundance of wickedness (Matthew 24:12; 1 Timothy 4; 2 Timothy 3, etc.), and the Gospel of the kingdom being preached to all nations (Matthew 24:13–14). As stated previously, all of these signs either have happened already or are currently in progress, but there are other signs, extremely specific events, that have not happened yet. For example, for the abomination of desolation to come to pass, as mentioned in Daniel 9:27, 11:31, 12:7–11; Matthew 24:15; 2 Thessalonians 2:3–4; and Revelation 13:14–18, the temple and temple sacrifices must first be reestablished by the Jews. Furthermore, for the Antichrist to sit in the temple of God and claim to be God, it's highly likely that he will sit on the mercyseat of the Ark of the Covenant, where God's spirit once sat (Exodus 25:22). This means that the mysterious Ark of the Covenant (made famous in the movie *Raiders of the Lost Ark*), which has been missing since Babylon invaded Israel over two thousand years ago, must be found and placed within the temple. In addition to these events that must transpire, all the national boundaries of Europe, a western portion of Asia, and a northern piece of Africa must be redrawn (Daniel 7:8, 7:24, 11:44—more on this in the next chapter).[5] Furthermore, the Antichrist will be revealed (2 Thessalonians 2:8), the Jews will flee into the wilderness in mass (Matthew 24:16), and there will be great signs in the heavens and on Earth (Matthew 24:4–31; Luke 21:11, 21:25–28; Acts 2:16–21; Revelation 6:12–17, 8:7–9:21, 11:1–13, 12:13–17, 13:1–8, 14:1–11, 16:1–21,

18:1–24). Some argue that the abomination of desolation was already fulfilled by Antiochus Epiphanies, who put an end to Jewish sacrifices and sacrificed a pig on the altar in the temple of the Lord. A close examination of *all* of the signs that must happen, however, reveals that this person was only a type of Antichrist who didn't quite fulfill all of Satan's plans. The simple fact is, Antiochus Epiphanies was not the Antichrist, because he's come and gone—and was not destroyed by Jesus Christ himself. This proves beyond a doubt that his actions, while they were an abomination, were not *the* abomination of desolation spoken of in prophecy.

- While the rapture of the church precedes the great tribulation, the second coming of Christ will occur at the conclusion of the great tribulation (Zechariah 14; Joel 3; Revelation 19:11–21, and many others mentioned above).
- Christ will obviously appear to all the inhabitants of Earth when he returns to establish his kingdom on Earth (Hebrews 9:28; Matthew 24:30, and many other scriptures which note God's obvious, visible kingdom on Earth).
- During the second advent of Christ, Christ will set foot on Earth when he establishes his kingdom (Daniel 2:44–45; 7:13–14, 7:18; Isaiah 9:6–7; Luke 1:32–33; Revelation 11:15).

One of the scriptures most commonly mistaken for a passage referring to the rapture of the church is the passage in Matthew 24:29–42. It is paralleled in Luke 17:34–37, where Jesus speaks of two in a field. One is taken while the other is left behind. So first we have a similarity to the rapture with the English word "taken." Second, Jesus stated in these passages that no one knows the day or hour of his return. This is still not a reference to the rapture of the church, however, simply because no one knows the day or the hour of the second coming of Christ. Preceding this discourse, Jesus starts out in Matthew 24:6 by clearly stating that certain things must happen before he returns, and this is clearly not the case with the rapture of the church (Philippians 3:20–21; Titus 2:11–13—early Christians believed the rapture of the Church could happen at any time). While no one knows the day or hour of Jesus's return, we do know that it won't happen until certain things happen first, so it is possible to know how *soon* it's going to be by watching for all of the signs Jesus gave. It's just not possible to know the exact day or hour.

The most telling reference to what being *taken* refers to in these two passages is found in Luke 17:37 and Matthew 24:27–28; Jesus's disciples asked where these people were being taken, and Jesus clearly answered their question: where the body is, there will the eagles also be gathered together. In other words, the ones taken away at the second advent will be those killed in the battle of Armageddon, who will make the carcasses for the eagles to eat, as pictured in Matthew 24:27–28 and Revelation 19:17–21. Those who are left will be the survivors of the battle of Armageddon, and they will be permitted to live on in the millennium, as in Zechariah 14:16–21 and Joel 2:29–32.

Luke 17:31–37 (bold emphasis added)

In that day, he which shall be upon the housetop, and his stuff in the house, let him not come down to take it away: and he that is in the field, let him likewise not return back. Remember Lot's wife. Whosoever shall seek to save his life shall lose it; and whosoever shall lose his life shall preserve it. I tell you, in that night there shall be two men in one bed; the one shall be taken, and the other shall be left. Two women shall be grinding together; the one shall be taken, and the other left. Two men shall be in the field; the one shall be taken, and the other left. **And they answered and said unto him, Where, Lord? And He said unto them, Wherever the body is, thither will the eagles be gathered together.**

Matthew 24:27–28 (bold emphasis added)

For as the lightning cometh out of the east, and shines even unto the west; so shall also the coming of the Son of man be. **For wherever the carcass is, there will the eagles be gathered together.**

Zechariah 14:16–21 (bold emphasis and bracketed comments added)

And it shall come to pass, that **every one that is left of all the nations which came against Jerusalem** [during the battle of Armageddon] shall even go up from year to year to worship the King, the LORD of hosts, and to keep the feast of tabernacles. And it shall be that whoso will not come up of all the families of the earth unto Jerusalem to worship the King, the LORD of hosts, even upon them shall be no rain. And if the family of Egypt go not up, and come not, that have no rain; there shall be the plague, wherewith the LORD will smite the heathen that come not up to keep the feast of tabernacles. This shall be the punishment of Egypt, and the punishment of all nations that come not up to keep the feast of tabernacles. In that day shall there be upon the bells of the horses, HOLINESS UNTO THE LORD; and the pots in the LORD's house shall be like the bowls before the altar. Yea, every pot in Jerusalem and in Judah shall be holiness unto the LORD of hosts: and all they that sacrifice shall come and take of them, and seethe therein: and in that day there shall be no more the Canaanite in the house of the LORD of hosts.

156 Aliens and the Antichrist

Joel 2:29–32 (bold emphasis added)

And also upon the servants and upon the handmaids in those days will I pour out my spirit. And I will show wonders in the heavens and in the earth, blood, and fire, and pillars of smoke. The sun shall be turned into darkness, and the moon into blood, before the great and the terrible day of the LORD come. And it shall come to pass, that **whosoever shall call on the name of the LORD shall be delivered: for in Mount Zion and in Jerusalem shall be deliverance**, as the LORD hath said, **and in the remnant whom the LORD shall call**.

2.2. More Than One Resurrection of the Dead, and More Than One Rapture

As mentioned earlier, two primary factors lead to the confusion as to when the rapture of the church will occur in relation to the tribulation. The first factor, which I have already discussed, is the common mistake of mixing up the rapture of the church with the second coming of Christ. I hope that I have provided enough information to distinguish these two events. The second leading factor contributing to the confusion of trying to figure out when the rapture of the church will occur is the fact that there is more than one rapture.

Concerning resurrections from the dead, the Bible speaks of a number of them (Elisha's bones—2 Kings 13:21, Lazarus—John 11:44, a little girl—Matthew 9:24–25), but some are distinguished with a special type of resurrection where the physical body is actually glorified and raised to eternal life with complete angelic immortality, then taken to heaven. This special type of resurrection is called a "rapture" (a term that stuck from the Latin vulgate "rapiemur," which in English is usually translated as "caught up").[6] Jesus was the first person to be raptured (1 Corinthians 15:23).

1 Corinthians 15:20–23 (bold emphasis added)

But now is **Christ rose from the dead, and become the first fruits of them that slept**. For since by man came death, by man came also the resurrection of the dead. For as in Adam all die, even so in Christ shall all be made alive. But every man in his own order: **Christ the first fruits; afterward they that are Christ's at his coming**.

Some might argue that Enoch (Genesis 5:24) and Elijah (2 Kings 2:11–12) were raptured before Jesus, but they were not translated into complete immortality, because they are the two witnesses spoken of in Revelation who will be killed by the beast that will ascend from the bottomless pit to make war against them (Revelation 11:7). This would be impossible to do if they were immortal like angels.

Revelation 11:7 (bold emphasis added)

And when they shall have finished their testimony, **the beast that will ascend out of the bottomless pit** shall make war against them, and **shall overcome them, and kill them.**

The first resurrection of the dead encompasses a total of five raptures, all pertaining to different groups of people. These groups are referred to as "orders" (1 Corinthians 15:23), and only *one* of them is the rapture of the church, even though the rapture of the tribulation saints will also be a "church" in the sense that it will be part of the body of Christ. Scripture simply distinguishes the Gentile church of this day and age from the tribulation saints of the future, because the prophetic emphasis in the days of the great tribulation will be focused on the nation of Israel and on the Jews in particular, rather than Christians throughout the world.

1 Corinthians 15:23

But every man in his own order: Christ the first fruits; afterward they that are Christ's at his coming.

Concerning the five raptures, they are listed as follows: First, the going to heaven of Christ and the saints that were resurrected from the dead immediately after Jesus was resurrected from the dead (Matthew 27:52; Ephesians 4:7–11; Acts 1:11). Second, the rapture of those who are Christ's at his coming—which is widely known as the "rapture of the church" (1 Corinthians 15:23; 1 Thessalonians 4:13–17). Third, the rapture of the 144,000 Jews in the middle of Daniel's seventieth week, under the seventh trumpet (Revelation 12:5, 14:1–5; Daniel 12:1; Isaiah 66:7–8). Fourth, the rapture of the two witnesses (Revelation 11:3–12). Fifth, the rapture of the tribulation saints who will be resurrected near the end of the great tribulation and right before the second coming of Christ (Acts 2:16–21, and also Revelation 20:4–6, with their spirits, but not necessarily their resurrected bodies, mentioned in Revelation 7:9–17, 15:2–4).

Matthew 27:52—first rapture (bold emphasis added)

And **the graves were opened; and many bodies of the saints which slept arose**, and came out of the graves after his resurrection, and went into the holy city, and appeared unto many.

1 Thessalonians 4:13–17—second rapture (rapture of the church) (bold emphasis added)

But I would not have you to be ignorant, brethren, concerning them which are asleep, that ye sorrow not, even as others which have no hope. For if we believe that Jesus died and rose again, even **so them also which sleep in Jesus will God bring with him**. For this we say unto you by the word of the Lord, that we which are alive and remain unto the coming of the Lord shall not prevent them which are asleep. For the Lord himself shall descend from heaven with a shout, with the voice of the archangel, and with the trump of God: and **the dead in Christ shall rise first: Then we which are alive and remain shall be caught up together with them in the clouds, to meet the Lord in the air**: and so shall we ever be with the Lord.

Revelation 12:5—third rapture* (bold emphasis added)

And she brought forth a man child, who was to rule all nations with a rod of iron: and **her child was caught up unto God**, and to his throne.

*For more information linking the "man child" with the 144,000 Jews, see God's Plan for Man.[7]

Revelation 11:3–12—fourth rapture (bold emphasis added)

And I will give power unto my two witnesses, and they shall prophesy a thousand two hundred and threescore days, clothed in sackcloth. These are the two olive trees, and the two candlesticks standing before the God of the earth. And if any man will hurt them, fire will proceed out of their mouth, and devour their enemies: and if any man will hurt them, he must in this manner be killed. These have power to shut heaven that it rain not in the days of their prophecy: and have power over waters to turn them to blood, and to smite the earth with all plagues, as often as they will. And when they shall have finished their testimony, the beast that will ascend out of the bottomless pit shall make war against them, and shall overcome them, and kill them. And their dead bodies shall lie in the street of the great city, which spiritually is called Sodom and Egypt, where also our Lord was crucified. And they of the people and kindreds and tongues and nations shall see their dead bodies three days and an half, and shall not suffer their dead bodies to be put in graves. And they that dwell upon the earth shall rejoice over them, and make merry, and shall send gifts one to another; because these two prophets tormented them that dwelt on the earth. **And after three days and an half the Spirit of life from God entered into them, and they stood upon their feet; and great fear fell upon them which saw them. And they heard a great voice from heaven saying unto them, Come up hither. And they ascended up to heaven in a cloud;** and their enemies beheld them.

Revelation 15:2–4—fifth rapture (bold emphasis added)

And I saw as it were a sea of glass mingled with fire: and **them that had gotten the victory over the beast, and over his image, and over his mark, and over the number of his name, stand on the sea of glass**, having the harps of God. And they sing the song of Moses the servant of God, and the song of the Lamb, saying, Great and marvelous are thy works, Lord God Almighty; just and true are thy ways, thou King of saints. Who shall not fear thee, O Lord, and glorify thy name? For thou only art holy: for all nations shall come and worship before thee; for thy judgments are made manifest.

The first (when Jesus was crucified), third (144,000 Jews), and fourth (two witnesses) raptures are easy to identify and differentiate from the second (the church) and fifth (tribulation saints) raptures. The common mistake is confusing the second and fifth raptures, because Christians will be killed during the entire period of the great tribulation, so many people assume that the rapture won't occur until the second coming of Jesus. This can't be accurate, however, because Jesus said it was possible to escape the tribulation (Luke 21:34–36). Nevertheless, why won't the tribulation saints escape the tribulation? Simple: because they won't believe until after the rapture of the church, after which they will believe, and many will then be martyred for their faith. In fact the rapture of the church will probably play a large role in the impetus of the newfound faith of these tribulation saints.

The old wives' tale about there being no hope after the rapture of the church has no scriptural support whatsoever. Nowhere in the Bible does it say anything that even hints that there is no hope for salvation during the great tribulation. Unbelief in the Gospel of truth will be widespread, but that doesn't mean that salvation won't be available. As long as we haven't died yet, and we haven't accepted the mark of the beast, his name, or the number of his name (Revelation 13:17; 20:4), and the day is not Judgment Day (Hebrews 9:27), then there is hope—and not a single scripture in the entire Bible can refute this fact.

Revelation 13:16–18

And he causes all, both small and great, rich and poor, free and bond, to receive a mark in their right hand, or in their foreheads: And that no man might buy or sell, save he that had the mark, or the name of the beast, or the number of his name. Here is wisdom. Let him that hath understanding count the number of the beast: for it is the number of a man; and his number is Six hundred threescore and six.

Revelation 20:4–6 (bold emphasis added)

And I saw thrones, and they sat upon them, and judgment was given unto them: and I saw the souls of them that were beheaded for the witness of Jesus, and for the word of God, and which had not worshipped the beast, neither his image, neither had received his mark upon their foreheads, or in their hands; and they lived and reigned with Christ a thousand years. But the rest of the dead lived not again until the thousand years were finished. This is the first resurrection. Blessed and holy is he that hath part in the first resurrection: on such the second death hath no power, but they shall be priests of God and of Christ, and shall reign with him a thousand years.

Hebrews 9:27

And as it is appointed unto men once to die, but after this the judgment…

In summation, to keep these scriptures from contradicting each other, as well as others I have yet to mention, a rapture must take place both before the great tribulation begins and near the end of the great tribulation, when it is almost complete. The rapture before the great tribulation applies to the *church* specifically, while the one that occurs at the conclusion of the great tribulation applies to those who accept Christ during the great tribulation.

In order to be fair, I welcome all readers to study the rebuttal to the pretribulation rapture of the church by going to http://www.graceonlinelibrary.org, and reading an article titled *Is the pretribulation rapture biblical?*[8] After reading this article, be sure to read all the comments posted beneath it. As I stated before, there is quite a bit of emotionalism in this article and in the comments that follow it, but I just ignore it and focus on the facts.

While having a bunch of contradictory views swimming around in one's head can be overwhelming, it's a healthy aspect of critical thinking. Back when I took a critical thinking course in college, I had to defend abortions, the legalization of drugs that are currently illegal, and argue against the application of the death penalty, because my professor made us argue against the views we believed in. He did this because one cannot sufficiently defend a position unless he or she is well versed in the opposing position. My textbook for the course, *Current Issues and*

Enduring Questions: A Guide to Critical Thinking and Argument, with Readings, was full of pros and cons on numerous controversial issues.

To apply a basic critical thinking technique to the topic of the pretribulation rapture of the church, one can simply make a list of pros and cons, and categorize each argument as weak or strong. Check to see if each argument and argument rebuttal is sufficiently answered. Analyze each argument individually, and then view all arguments for a particular case together as a collective whole. In many cases, taking a position on a particular subject is a simple mathematical process.

My critical thinking course helped me a great deal in researching topics and developing conclusions after studying both sides of an argument. None of the students in my class ever knew what positions our teacher believed in, except for one. He said he believed in the death penalty, because he had a personal list of people he wished would receive it. Of course, he was joking, *I think*.

Back to the topic at hand, concerning the rapture that will take place near the end of the great tribulation, exactly when will the tribulation saints be raptured? Revelation 6:9–11 states that many tribulation saints will be martyred before the fifth seal, which is during the first three and one-half years of the Antichrist. This period is expected to occur almost immediately after the rapture of the church. Certainly Satan won't waste time after the restrainer (the church) is removed from Earth (2 Thessalonians 2:6–8).

> Revelation 6:9–11 (bold emphasis added)
>
> And when he had opened the fifth seal, **I saw under the altar the souls of them that were slain for the word of God,** and for the testimony which they held: And they cried with a loud voice, saying, How long, O Lord, holy and true, dost thou not judge and avenge our blood on them that dwell on the earth? And white robes were given unto every one of them; and it was said unto them, that they should rest yet for a little season, **until their fellow servants also and their brethren, that should be killed as they were, should be fulfilled**.

The saints who are killed during this first three and one-half years are told to rest until the remaining tribulation saints are killed in the second half of the great tribulation (Revelation 14:13). Then vengeance will be taken out on those who will kill them.

Revelation 14:13

> And I heard a voice from heaven saying unto me, Write, Blessed are the dead who die in the Lord from henceforth: Yea, says the Spirit that they may rest from their labors; and their works do follow them.

The tribulation saints, therefore, will be ruthlessly slaughtered during the entire seven-year reign of terror of the Antichrist. Near the very tail end of the great tribulation, the saints who were killed will be resurrected. Then the saints who are still alive at that time, along with those who will be resurrected, will all be *taken up* to heaven. There, they will be present for the marriage supper of the Lamb. Revelation 19:1–9 states that the wife is ready, and the "wife," the city of God's people in heaven, could not be ready for the marriage supper of the Lamb, if some of them are still stuck down on Earth and might not even believe in Jesus yet. Being ready means everyone who will ever believe in Jesus must be present and accounted for. The rapture of these tribulation saints, therefore, will occur right at the tail end of the great tribulation, but before the return of Christ to Earth. In fact, shortly following this marriage supper of the Lamb, everyone will saddle up on their white horses (perhaps awaiting them in the first heaven) and ride off to make war at the second coming of Christ (Revelation 19:14–21).

Revelation 19:1–9 (bold emphasis added)

> And after these things I heard a great voice of much people in heaven, saying, Alleluia; Salvation, and glory, and honor, and power, unto the Lord our God: For true and righteous are his judgments: for He hath judged the great whore, which did corrupt the earth with her fornication, and hath avenged the blood of his servants at her hand. And again they said, Alleluia. And her smoke rose up for ever and ever. And the four and twenty elders and the four beasts fell down and worshipped God that sat on the throne, saying, Amen; Alleluia. And a voice came out of the throne, saying, Praise our God, all ye his servants, and ye that fear him, both small and great. And I heard as it were the voice of a great multitude, and as the voice of many waters, and as the voice of mighty thunderings, saying, Alleluia: for the Lord God omnipotent reigns. Let us be glad and rejoice, and give honor to him: **for the marriage of the Lamb is come and his wife hath made herself ready**. And to her was granted that she should be arrayed in fine linen, clean and white: for the fine linen is the righteousness of saints. And he said unto me, Write, Blessed are they which are called unto the marriage supper of the Lamb. And he said unto me, These are the true sayings of God.

Revelation 19:11–21 (bold emphasis added)

And I saw heaven opened, and behold a white horse; and He that sat upon him was called Faithful and True, and in righteousness He doth judge and make war. His eyes were as a flame of fire, and on his head were many crowns; and He had a name written, that no man knew, but Him. And He was clothed with a vesture dipped in blood: and his name is called The Word of God. **And the armies which were in heaven followed him upon white horses, clothed in fine linen, white and clean.** And out of his mouth goes a sharp sword, that with it He should smite the nations: and He shall rule them with a rod of iron: and He treads the winepress of the fierceness and wrath of Almighty God. And He hath on his vesture and on his thigh a name written, KING OF KINGS, AND LORD OF LORDS. And I saw an angel standing in the sun; and He cried with a loud voice, saying to all the fowls that fly in the midst of heaven, Come and gather yourselves together unto the supper of the great God; That ye may eat the flesh of kings, and the flesh of captains, and the flesh of mighty men, and the flesh of horses, and of them that sit on them, and the flesh of all men, both free and bond, both small and great. **And I saw the beast, and the kings of the earth, and their armies, gathered together to make war against him that sat on the horse, and against his army. And the beast was taken, and with him the false prophet that wrought miracles before him, with which he deceived them that had received the mark of the beast, and them that worshipped his image. These both were cast alive into a lake of fire burning with brimstone. And the remnant were slain with the sword of him that sat upon the horse, which sword proceeded out of his mouth**: and all the fowls were filled with their flesh.

2.3. Who Hinders the Antichrist?

I've already disclosed that I believe that the rapture of the church will occur before the great tribulation, but believe it or not, I haven't discussed all of the details supporting this conclusion. For example, I referenced 2 Thessalonians 2:6–8; Revelation 1:19, 4:1 as scriptures that prove the rapture of the church will occur before the great tribulation, but I didn't elaborate on how these scriptures tie together to meet this conclusion.

Starting with 2 Thessalonians 2:6–8, note that someone must be taken out of the way before the Antichrist is revealed, and that someone (he) is referred to as a withholder (hinderer) of the Antichrist. Who could this be?

2 Thessalonians 2:6–8 (bold emphasis and bracketed comments added)

And now ye know what **withholds** that he might be revealed in his time. For the mystery of iniquity doth already work: only he who now letteth will let, until **he [the church] be taken out of the way. And then shall that Wicked [the Antichrist] be revealed**, whom the Lord shall consume with the spirit of his mouth, and shall destroy with the brightness of his coming.

Some Christians say that this hinderer is the Holy Spirit, but this can't be accurate, because the Holy Spirit is never taken out of the way of anything, much less the Antichrist. In fact the Holy Spirit is actually visibly active during the great tribulation, when the Antichrist will be wreaking havoc.

Others have postulated that this hinderer refers to government, but the Antichrist will be a mastermind at ruling and manipulating governments all through his reign of terror, so this can't be right either. The only likely candidate this "he" can be is the church, who is a "he" in the sense of being the body of Christ. The church makes perfect sense, because the fervent prayers of the righteous have an extremely powerful effect (James 5:16). Suddenly removing all of the Christians from Earth will be the removal of every single voice against governmental authorities in all matters of righteousness, justice, and morality—everything with spiritual significance. The implications of such would be staggering; the world will be a spiritual vacuum, creating an oasis for any charismatic dictator to step forward and assume control. Many Christians think they have no power, voice, or effect in the world, but they couldn't be more wrong. For those Christians who are inactive and ineffective—consider this a call to action!

A further clue pointing to the restrainer of the Antichrist being the church is found by analyzing references to the church in the book of Revelation. The book of Revelation is divided into three parts, and these parts are outlined in Revelation 1:19.

Revelation 1:19 (bold emphasis added)

Write the **things which thou hast seen**, and **the things which are**, and the **things which shall be hereafter**...

First, "the things which thou hast seen" are the things in the vision of Christ in Revelation 1. Second, "the things which are" are the things concerning the churches, covered in Revelation 2–3, which John himself could see were in the world in his day. Third, "the things which shall be hereafter" refer to the time that comes after the churches.

I've read Revelation a million times and never noticed this clear outline until Finis Dake pointed it out to me. Now that I know about it, I don't know why I never saw it before, because it's obvious. Everything in chapter 1 of Revelation is about the vision of Christ; everything about the churches is listed in chapters 2 and 3; and everything after the churches comes after Revelation 4:1.

Revelation 4:1 (bold emphasis added)

After this I looked, and, behold, **a door was opened in heaven**: and the first voice which I heard was as it were of a trumpet talking with me; which said, **Come up hither, and I will show thee things which must be hereafter.**

The transition that takes place in Revelation 4:1 is significant because it reveals the rapture of the church in two ways. First, before Revelation 4:1, the church is mentioned eighteen times, but then after Revelation 4:1, the church isn't mentioned anymore until the very tail end of the last chapter of Revelation, after the revelation is complete—and Jesus is simply telling John that the revelation was given to him to give to the churches.

Revelation 22:16

I Jesus have sent mine angel to testify unto you these things in the churches. I am the root and the offspring of David, and the bright and morning star.

The second way that everything after the churches points to the rapture of the church is indirectly, by the fact that John himself is taken up to heaven at the exact point in time when the rapture of the church will happen—after the churches, in Revelation 4:1. As the revelation unfolds, John is hearing messages for the churches in chapter 3, and then suddenly after these things—that is, after the churches—John is caught up to heaven to see the rest of the vision unfold.

2.4. What Signs Point to the Nearness of the Rapture of the Church?

While there are no signs that must occur before the rapture of the church, it is known that the rapture of the church occurs approximately seven years before the second coming of Christ (Daniel 9:27—each day represents one year in the seventieth week of Daniel's vision, and all weeks but the seventieth week have been fulfilled). The nearness of the rapture of the church can therefore be gauged by looking for signs that must happen before the second coming of Christ. However close we are to the second coming of Christ, we are seven years closer to the rapture of the church.

Most of the signs that must occur before the second coming of Christ have already occurred, though their relevance is more noticeable these days. Wars and rumors of wars, for example, constitute a somewhat vague prophecy that could apply to any time in Earth's history. However, there have been more wars fought and more people killed in the past hundred years than all of the wars fought in the rest of all known history combined.[9] Wars and rumors of wars, therefore, is a sign that bears unique relevance in this generation.

What about the increase of knowledge and traveling in the end times that the prophet Daniel prophesied?

> Daniel 12:4 (bold emphasis added)
>
> But thou, O Daniel, shut up the words, and seal the book, even to the time of the end: **many shall run to and fro, and knowledge shall be increased**.

Does not this current generation see more travel than any generation preceding it? Has not this current generation increased in knowledge (especially with the advent of computers, the Internet, and advanced communications systems), more than any generation preceding it? These things are not only increasing; they are also accelerating at an exponential rate unlike ever before in the history of humanity. Again this prophecy in Daniel bears unique relevance in this current generation, more than any generation prior.

As for specific prophecies that must happen, when Israel became a nation again in 1948, the entire world was shocked. Nothing could have been more unlikely prior to the nineteenth century, during the days of the Ottoman Empire. The nation of Israel was an ancient legend passed down from Jewish elders—an almost mythical past of the Jews. As for the Hebrew language, it was a relic of antiquity; yet within a few scant years, this entire nation rose from the dust of a forsaken land, and the Hebrew language was revived again to become the national language of the Jewish people.[10] That's about as likely as NATO declaring the Cherokee Native American tribe to be the new rulers of America, and everyone in America learning to speak Cherokee—and then it actually happens!

The Jews were the underdogs of the entire world. Right after the Holocaust, which claimed the lives of over six million Jews, the British mandate over the region of Israel-Palestine ended, and the Jews declared Israel to be an independent nation again, in accordance with the established boundaries of the British partition plan.[11] While the rest of the world silently concurred with Israel's declaration, primarily out of sympathy for the anti-Semitism the Jews endured, the nations surrounding Israel didn't share in that same sympathy. No sooner had Israel been declared a nation did all the countries surrounding Israel (Egypt, Syria, Transjordan, Lebanon, Iraq, and Arabia), declare war against the Jews—Holocaust or no Holocaust.[12]

Apparently the sentiment of approval expressed to the Jews at that time from the global community didn't mean much because the Jews didn't receive much help in defending themselves. As if the Holocaust weren't bad enough, now the Jews had to fight for their parcel of territory to call home—a battle against a half-dozen armies, with clear superiority in heavy arms and firepower.[13] (I only referenced a few sources for this information, but these facts can easily be obtained from literally hundreds of Web sites, books, documentaries, and the like.)

Some say the Jews can fight, and certainly there is evidence to support this fact, but I strongly believe God was (and still is) on their side because they whipped the pants off everyone, and they've been fighting ever since. Every time they're invaded, they fight back with such ferocity that they usually gain more territory, but then later give it back after a great deal of disputing. The Middle East is a hotbed of activity surrounding Israel. As I was saying before, this small country becoming a nation again has prophecy written all over it. The global war on terror being waged right now stems primarily from contention over the Jews occupying their original homeland of Israel.

With the establishment of Israel as a nation again, it's now more likely that the Jewish temple will be restored and the temple sacrifices reestablished, both of which are very specific signs that throughout history were extremely unlikely—yet they are now much closer to being a reality. Anyone reading up on this will see that there is activity brewing.[14]

As for the Ark of the Covenant, there are a number of leads as to its location (see http://www.wyattmuseum.com and http://www.christiananswers.net).[15] Also, the birth pangs of the Earth increase with each passing year, just as best-selling author Hal Lindsey illustrates in depth in his many Bible prophecy books. The weather is going insane on a global scale, with one devastating natural disaster after another. The December 2004 tsunami that resulted from the Sumatra-Andaman earthquake, took over 280,000 lives, making it one of the deadliest disasters in modern history.[16] This earthquake was so powerful, it literally liquefied the ocean floor where it struck, forming a completely different terrain. Another huge earthquake struck in Pakistan just ten months later, in October 2005, which resulted in 80,000+ deaths.[17] The entire hurricane season for year 2005 shattered all previous hurricane season records, including the most powerful, the most costly, and the greatest number of hurricanes ever to strike in a single year.[18] Might these extreme weather patterns be telling us something?

These recent events make even the secular community wonder if the end of the world is near. For every sign that points to the second coming of Christ, there are a number of reports on events currently taking place that reveal a progression of the fulfillment of said prophecies.

We are very close. If I don't die prematurely, I doubt I will die at all.

5

The Rise of the Antichrist

Shortly following the rapture of the church, while people are still confused, frightened, and otherwise freaked out (though many will come to the realization of what just happened), Earth will be ripe for a massive satanic invasion, because the church, which has been holding back the emergence of the Antichrist, will be removed (2 Thessalonians 2:6–8). What will happen next?

Satan has diligently been trying to take over the governments of the world ever since Adam. The pharaoh of Egypt (Exodus), the king of Assyria—Antiochus Epiphanies (Isaiah 10:5), the king of Tyre (Ezekiel 28:11–19), the king of the Grecian Empire—Alexander the Great (Daniel 8:5–26), and Roman rulers such as Herod, his son with the same name, and Nero (Matthew 2; Luke 1:5, 23:6–15; Acts 25:11; 1 Timothy 2:2), were all heavily influenced by the power of Satan. Other national leaders from more recent times, such as Adolph Hitler, Saddam Hussein, and the like are all prime examples of Satan attempting to take over the world. All of these rulers were satanically empowered; most of them were bent on the destruction of the Jews in particular. Most of them had aspirations for world domination, and in most cases, they demanded worship as deities. They were all *types* of antichrists, but they weren't *the* Antichrist.

After the rapture of the church, Satan will finally be able to make his big move, empowering a new national leader with more cunning than ever before. Even non-Christians sometimes wonder with trepidation about this figure—the future Antichrist. I suppose it's easy for some people to mistake Jesus's mercy for nonthreatening pacifism, despite the fact that the danger he poses to dissidents is far greater than anything the Antichrist will ever dish out. Regardless of this fact, however, the term "Antichrist" strikes fear into the hearts of many. This man will be Satan's masterpiece; but before the Antichrist rises to prominence, a few things must transpire. Aside from the rapture of the church, which I've already covered, a number of major governmental changes must also take place. Some of these events might even occur before the rapture of the church. These changes will be centralized in the region surrounding the Mediterranean Sea, in the area known as the ancient Roman Empire.

1. Big Changes Around the Mediterranean Sea

The book of Daniel was written about 530 BC, yet predicted the fall of the Babylonian Empire, as well as the rise and fall of the Medo-Persian Empire, the Grecian Empire, and the Roman Empire, all of which were located around the Mediterranean Sea. Proponents of prophecy have tried to explain this accuracy by suggesting the book was written later in the second century, but this still doesn't preclude all the prophecies it predicts. The rise of the Roman Empire is clearly depicted, and the coming of the "Anointed One, the ruler," 483 years after the issuing of the decree to restore and rebuild Jerusalem (Daniel 9:25) works out to the time of Jesus's ministry. Both of these prophecies occurred after the second century, so a second-century date still doesn't preclude the prophetic elements of the book of Daniel.[1]

> Daniel 9:25 (bold emphasis added)
>
> Know therefore and understand, that **from the going forth of the commandment to restore and to build Jerusalem unto the Messiah the Prince shall be seven weeks, and threescore and two weeks**: the street shall be built again, and the wall, even in troublous times.

Until recently, I haven't seriously undertaken the task of attempting to understand the visions of Daniel or John. They're so symbolic, I figured they could mean practically anything. I soon discovered this was a flawed conception on my part, because after only a few days of conducting research into these prophecies, I realized that most scholars agree on about 90 percent of the interpretations. For example, both the visions of Daniel and John speak of strange animals with horns. Most everyone agrees that these creatures represent nations and the rulers associated with them. The fact that the scriptures themselves state this verbatim (Daniel 7:17, 7:23–24, 8:20–23; Revelation 13:1–18, which is clarified by Revelation 17:8–17), and the fact that the historical record is in exact agreement with these prophecies that have already been fulfilled makes this assessment obvious. The majority of the visions of Daniel, for example, have already been fulfilled, so who can argue against the interpretation? Simply follow the pattern, which is about 75 percent complete, and one can clearly see the events that will transpire in the future.

Daniel 7:17

These great beasts, which are four, are four kings, which shall arise out of the earth.

Daniel 7:23-24

Thus he said, The fourth beast shall be the fourth kingdom upon earth, which shall be diverse from all kingdoms, and shall devour the whole earth, and shall tread it down, and break it in pieces. And the ten horns out of this kingdom are ten kings that shall arise: and another shall rise after them; and he shall be diverse from the first, and he shall subdue three kings.

Daniel 8:20-23

The ram which thou sawest having two horns are the kings of Media and Persia. And the rough goat is the king of Grecia: and the great horn that is between his eyes is the first king. Now that being broken, whereas four stood up for it, four kingdoms shall stand up out of the nation, but not in his power. And in the latter time of their kingdom, when the transgressors are come to the full, a king of fierce countenance, and understanding dark sentences, shall stand up. The beast that thou sawest was, and is not; and shall ascend out of the bottomless pit, and go into perdition: and they that dwell on the earth shall wonder, whose names were not written in the book of life from the foundation of the world, when they behold the beast that was, and is not, and yet is.

Revelation 17:9-17

And here is the mind which hath wisdom. The seven heads are seven mountains, on which the woman sits. And there are seven kings: five are fallen, and one is, and the other is not yet come; and when he cometh, he must continue a short space. And the beast that was, and is not, even he is the eighth, and is of the seven, and will go into perdition. And the ten horns which thou sawest are ten kings, which have received no kingdom as yet; but receive power as kings one hour with the beast. These have one mind, and shall give their power and strength unto the beast. These shall make war with the Lamb, and the Lamb shall overcome them: for He is Lord of lords, and King of kings: and they that are with him are called, and chosen, and faithful. And he said unto me, The waters which thou sawest, where the whore sits, are peoples, and multitudes, and nations, and tongues. And the ten horns which thou sawest upon the beast, these shall hate the whore, and shall make her desolate and naked, and shall eat her flesh, and burn her with fire. For God hath put in their hearts to fulfill his will, and to agree, and give their kingdom unto the beast, until the words of God shall be fulfilled.

Concerning the interpretational aberrations deriving from these prophecies, some people refuse to acknowledge that Daniel is speaking about a specific geographical region controlled under governmental authorities. For example, some people try to tie the United States into the picture by saying that the leopard in one of Daniel's visions represents Germany rather than the ancient Grecian Empire.[2] The national symbol of Germany is a leopard (though it also uses an eagle), and Germany was divided by four countries following WWII (United States, England, France, and Russia)—but Germany wasn't actually split up four ways, as the Grecian Empire was.

The United Nations is frequently read into scripture as well, but the United Nations includes nations that are nowhere near the area surrounding the Mediterranean Sea. Numerous different interpretations of Daniel's prophecies have been proposed, but as far as I can tell from the research I've conducted, they conclude with facts that are not consistent with scripture. For example, saying the Antichrist can come from America doesn't fit with numerous key scriptures (which I will soon divulge). Furthermore, proponents of the United Nations theories frequently falsely conclude that the Antichrist will be a global dictator, but scripture is very clear about the fact that the Antichrist will never be a worldwide dictator; he will be at war for the majority of his seven-year reign. How can he be at war with his own kingdom? Also, many people will never receive the mark of the beast, yet they will survive to see the millennium. The Antichrist, therefore, will not have the entire world completely under his control. It won't be until the very end of his seven-year reign that the entire world will align under his military command to converge on Israel, but being a military commander isn't the same as being a dictator over the entire planet. I suspect the Antichrist will be a military leader in something that resembles the United Nations, but this is very different from being an actual global dictator.

One of the primary reasons people are analyzing Daniel's prophecies and seeing these alternate interpretations is because they want to identify the nation out of which the Antichrist will arise. This can only be done by harmonizing all of the scriptures that speak about him, as I will clearly demonstrate in successive sections. The Antichrist can only come from one specific place, and the Bible points out exactly where that place is; there is no need for a guessing game.

Daniel's visions actually aren't that hard to understand; all it takes is a little common sense to see that they talk about ten kingdoms in a specific geographical region (within the vicinity of the Mediterranean Sea). Saying the ten kingdoms of Daniel's visions are ten governmental types, or an economic treaty between ten countries, or any other manner of creative interpretation, goes further and further away from what scripture actually says.

1.1. The Formation of the Revised Roman Empire

Concerning the prophecies in the book of Daniel, as well as similar prophecies covering the same events given by the apostle John while exiled on the Isle of Patmos, a new revision of the ancient Roman Empire will be formed in the last days. This empire will be the seventh world-dominating empire mentioned in scripture. The first world-dominating empire was Egypt, the second Assyria, and then the four kingdoms mentioned above, which are Babylon, Medo-Persia, the Grecian Empire, and the Roman Empire (see the visions of Daniel in Daniel 7:1–12:13, and visions of John in Revelation, as expounded on in *God's Plan for Man*, by Finis Dake, for more detailed information). For anyone wishing to delve into the fifty-billion-page version of the brief exposition I outline here, I actually recommend reading all one thousand pages of *God's Plan for Man*, by Finis Dake. Suffice it to say, my version of these visions is the "Cliffs Notes" version because I have other things to write about rather than driveling on about every detail. The Internet is full of this same information as well. Almost all students of the Bible agree about the identity of the kingdoms of the world leading up to the revised Roman Empire.

The new *revised* Roman Empire will be a conglomeration of ten kingdoms, ruled by ten kings—so it isn't a *revived* Roman Empire, as some teach because there will be ten kings rather than one. (The original Roman Empire had only one king, and to suggest that this Roman Empire will be revived rather than revised would suggest that it will have one king—and it never will, not even when the Antichrist rises to power.)

This revised Roman Empire will encompass a very large tract of land that once housed fifty-four provinces.[3] There are now (as of the year 2006) over twenty complete countries within the original territory of the ancient Roman Empire, and about twenty more countries if one counts areas where only portions of countries are included.[4] All of these countries surround the Mediterranean Sea and span three continents, which includes most of Europe, a western portion of Asia, and a northern portion of Africa.

Now what would have to take place that would shuffle national boundaries around to such a degree that nearly forty countries become ten? An easy answer would be war—and a very large one at that. Keep in mind that this *has* to happen before the Antichrist will rise to power.

Not many people foresaw the collapse of communism, which fell without a single shot fired. Because of this unforeseeable event, for all we know, groups of countries could simply unite peacefully (meaning two or more countries erasing their borders and actually merging into a single country), though I know of no time in history where this has ever happened. War is always the catalyst that defines national boundaries. Most of the nations in this region are now democratic, which at least makes a peaceful transition more plausible than it was in the days of Communist rule, but a peaceful transition still looks unlikely. Simply look at what's going on over there right now! The trend of the region has always been war, especially in the Syrian quadrant of the future empire in question. I think the most likely scenario will be the backlash of Islam against the coalition that defeated Iraq. I wouldn't be surprised if Syria and its predominantly Muslim neighbors go to war with each other over the problems with terrorism and the introduction of Western democracy, and then conglomerate into larger countries. They might even decide to unite agreeably against Western democracy, but I doubt it.

Once they're united, however that comes about, they'll push their attacks west into Europe, which will force the European countries to unite in order to fight back. Whatever the scenario, the revised Roman Empire must exist as a ten-kingdom entity before the Antichrist rises to power.

1.2. The First Conquest of the Antichrist

Once the ten kingdoms are in place, the Antichrist is identified by both the prophet Daniel and the apostle John as being one of the kings of one of these nations. Daniel in particular was given three visions that speak of which area the Antichrist will rise to power from, and each vision narrows down his location with more precision.

First, the Antichrist is mentioned as being one of the kings of the ten kingdoms of the revised Roman Empire (ten toes of beast in Daniel 2; ten horns of beast in Daniel 7:23–24; ten horns of beast in Revelation 12:3; 13:1–4; 17:8–17). In Daniel's second vision, the Antichrist is said to arise out of the area that was once part of the ancient Grecian Empire (Daniel 8:8–9, 8:21–23). This region overlaps the eastern portion of the ancient Roman Empire; it is symbolized by the metals of the image in Daniel 2, the four beasts of Daniel 7, and the he-goat of Daniel 8:5–26. In Daniel's third vision, it was shown that the Grecian Empire would be divided into four parts following the death of Alexander the Great (Daniel 8:8–9, 8:21–23, 11:4), and then Daniel saw which of these four parts the Antichrist would arise from. The quadrant he will come from will be the Syrian quadrant. This can be determined from the fact that he will be in the nation directly north of Egypt (the Antichrist is mentioned as being the king to the north of Egypt—Daniel 11:36–46), which completely rules out a number of countries different theorists have speculated through the years, such as the Vatican, Germany, England, and the United States. This area today is the location of modern Syria, Lebanon, Israel, Jordan, Saudi Arabia, Iraq, and Iran.

Daniel 7:23–24* (bold emphasis added)

Thus he said the fourth beast shall be the fourth kingdom upon earth, which shall be diverse from all kingdoms, and shall devour the whole earth, and shall tread it down, and break it in pieces. And **the ten horns out of this kingdom are ten kings that shall arise: and another shall rise after them**; and he shall be diverse from the first, and he shall subdue three kings.

The Antichrist is depicted as arising out of one of the ten kings of the revised Roman Empire.

Daniel 8:21–23* (bold emphasis and bracketed comments added)

The ram which thou sawest having two horns are the kings of Media and Persia. And the rough goat is the king of Grecia: and the great horn that is between his eyes is the first king [Alexander the Great]. Now that being broken, whereas four stood up for it, **four kingdoms shall stand up out of the nation**, but not in his power. **And in the latter time of their kingdom, when the transgressors are come to the full, a king of fierce countenance, and understanding dark sentences [the Antichrist], shall stand up.** The beast that thou sawest was, and is not; and shall ascend out of the bottomless pit, and go into perdition: and they that dwell on the earth shall wonder, whose names were not written in the book of life from the foundation of the world, when they behold the beast that was, and is not, and yet is.

Clearly the Antichrist is mentioned as coming from one of the four quadrants of the ancient Grecian Empire.

Daniel 11:36–46* (bold emphasis and bracketed comments added)

And the king shall do according to his will; and he shall exalt himself, and magnify himself above every god, and shall speak marvelous things against the God of gods, and shall prosper till the indignation be accomplished: for that that is determined shall be done. Neither shall he regard the God of his fathers, nor the desire of women, nor regard any god: for he shall magnify himself above all. But in his estate shall he honor the God of forces: and a god whom his fathers knew not shall he honor with gold, and silver, and with precious stones, and pleasant things. Thus shall he do in the most strong holds with a strange god, whom he shall acknowledge and increase with glory: and he shall cause them to rule over many, and shall divide the land for gain. **And at the time of the end shall the king of the south [Egypt, shown below] push at him: and the king of the north shall come against him like a whirlwind, with chariots, and with horsemen, and with many ships; and he shall enter into the countries, and shall overflow and pass over.** He shall enter also into the glorious land, and many countries shall be overthrown: but these shall escape out of his hand, even Edom, and Moab, and the chief of the children of Ammon. He shall stretch forth his hand also upon the countries: and **the land of Egypt shall not escape. But he shall have power over the treasures of gold and of silver, and over all the precious things of Egypt**: and the Libyans and the Ethiopians shall be at his steps. **But tidings out of the east [China] and out of the north [the former Soviet Union] shall trouble him: therefore he shall go forth with great fury to destroy and utterly to make away many.** And he shall plant the tabernacles of his palace between the seas in the glorious holy mountain; yet he shall come to his end, and none shall help him.

A nation north of Egypt yet south of another great world empire (which must be the former Soviet Union) would place the Antichrist in the Syrian quadrant of the ancient Grecian Empire, which will also be one of the eastern ten nations of the revised Roman Empire.

Aside from the facts listed above, the Antichrist is also called the Assyrian in scripture (Isaiah 10:20–27, 30:18–33, 31:4–32:20; Micah 5:3–15), which further identifies him with the nation of Syria.

Isaiah 10:20–27 (bold emphasis and bracketed comments added)

And it shall come to pass in that day, that the remnant of Israel, and such as are escaped of the house of Jacob, shall no more again stay upon him that smote them; but shall stay upon the LORD, the Holy One of Israel, in truth. The remnant shall return, even the remnant of Jacob, unto the mighty God. For though thy people Israel be as the sand of the sea, yet a remnant of them shall return: the consumption decreed shall overflow with righteousness. For the Lord GOD of hosts shall make a consumption, even determined, in the midst of all the land. Therefore thus says the Lord GOD of hosts, O my people that dwell in Zion, be not afraid of **the Assyrian: he [the Antichrist]** shall smite thee with a rod, and shall lift up his staff against thee, after the manner of Egypt. For yet a very little while, and the indignation shall cease, and mine anger in their destruction. And the LORD of hosts shall stir up a scourge for him according to the slaughter of Midian at the rock of Oreb: and as his rod was upon the sea, so shall he lift it up after the manner of Egypt. And it shall come to pass in that day, that his burden shall be taken away from off thy shoulder, and his yoke from off thy neck and the yoke shall be destroyed because of the anointing.

Micah 5:3-6 (bold emphasis and bracketed comments added)

Therefore will he give them up, until the time that she which travails hath brought forth: then the remnant of his brethren shall return unto the children of Israel. And he shall stand and feed in the strength of the LORD, in the majesty of the name of the LORD his God; and they shall abide: for now shall he be great unto the ends of the earth. And this man shall be the peace, when **the Assyrian [the Antichrist] shall come into our land: and when he shall tread in our palaces,** then shall we raise against him seven shepherds, and eight principal men. And they shall waste the land of Assyria with the sword, and the land of Nimrod in the entrances thereof: thus shall he deliver us from the Assyrian, when he cometh into our land, and when he treads within our borders.

Obviously from the facts mentioned above, the Antichrist will arise from the land known as ancient Assyria; this fact can clearly be determined by the title of the Antichrist (the Assyrian) and his geographical region—north of Egypt, south of another major world power, the former Soviet Union, and west of yet another major world power, China. Modern-day Syria encompasses the majority of the Assyrian quadrant, which was established under the ancient Grecian Empire, but Lebanon is also a candidate for being the nation where the Antichrist will initially rise to power because it too is north of Egypt, and it was once within the borders of ancient Assyria. Notably it's also a very small nation. Another name for the Antichrist is "little horn" (Daniel 7:8, 7:24, 8:9, 8:23), which denotes the possibility that a small nation is being spoken of, in comparison to the other nations around it. Lebanon is among the smallest nations to the north of Egypt. Whichever of the ten nations the Antichrist comes from will be apparent soon enough, however, because the formation of the revised Roman Empire won't be composed of ten nations for very long. The Antichrist will immediately begin his first conquest and go to war with his neighbors. Among these neighbors will certainly be Iraq because the ancient city of Babylon is located within Iraq, and that's the first place where he will establish his capital in his newly conquered territory (Isaiah 14:1-11; Zechariah 5:5-11; Revelation 16:17-21; 18:1-24).

Isaiah 14:1 (bold emphasis and bracketed comments added)

For the LORD will have mercy on Jacob, and will yet choose Israel, and set them in their own land: and the strangers shall be joined with them, and they shall cleave to the house of Jacob. And the people shall take them, and bring them to their place: and the house of Israel shall possess them in the land of the LORD for servants and handmaids: and they shall take them captives, whose captives they were; and they shall rule over their oppressors. And it shall come to pass in the day that the LORD shall give thee rest from thy sorrow, and from thy fear, and from the hard bondage wherein thou wast made to serve, that thou shalt take up this proverb against the **king of Babylon [the Antichrist]**, and say, How hath the oppressor ceased! The golden city ceased!

Zechariah 5:8–10 (bold emphasis and bracketed comments added)

And he said, This is wickedness. And he cast it into the midst of the ephah; and he cast the weight of lead upon the mouth thereof. Then lifted I up mine eyes, and looked, and, behold, there came out two women, and the wind was in their wings; for they had wings like the wings of a stork: and they lifted up the ephah between the earth and the heaven. Then said I to the angel that talked with me, Whither do these bear the ephah? And he said unto me, To build a house in **the land of Shinar [land of Babylon]**: and it shall be established, and set there upon her own base.

Revelation 16:16–21 (bold emphasis added)

And he gathered them together into a place called in the Hebrew tongue Armageddon. And the seventh angel poured out his vial into the air; and there came a great voice out of the temple of heaven, from the throne, saying, It is done. And there were voices, and thunder, and lightning; and there was a great earthquake, such as was not since men were upon the earth, so mighty an earthquake, and so great. And **the great city** was divided into three parts, and the cities of the nations fell: and **great Babylon** came in remembrance before God, to give unto her the cup of the wine of the fierceness of his wrath. And every island fled away, and the mountains were not found. And there fell upon men a great hail out of heaven, every stone about the weight of a talent: and men blasphemed God because of the plague of the hail; for the plague thereof was exceeding great.

Babylon, the capital city of the Antichrist kingdom, will be moved during the middle of Daniel's seventieth week, after the first three and one-half years of the reign of the Antichrist. It will be moved to Jerusalem (Daniel 9:27, 11:40–45; 2 Thessalonians 2:3–4; Revelation 13).

Daniel 9:27 (bold emphasis added)

And he shall confirm the covenant with many for one week: and **in the midst of the week he shall cause the sacrifice and the oblation to cease**, and for the overspreading of abominations he shall make it desolate, even until the consummation, and that determined shall be poured upon the desolate.

Daniel 11:40–45 (bold emphasis and bracketed comments added)

And at the time of the end shall the king of the south push at him: and the king of the north shall come against him like a whirlwind, with chariots, and with horsemen, and with many ships; and he shall enter into the countries, and shall overflow and pass over. **He shall enter also into the glorious land, [Israel]** and many countries shall be overthrown: but these shall escape out of his hand, even Edom, and Moab, and the chief of the children of Ammon. He shall stretch forth his hand also upon the countries: and the land of Egypt shall not escape. But he shall have power over the treasures of gold and of silver, and over all the precious things of Egypt: and the Libyans and the Ethiopians shall be at his steps. But tidings out of the east and out of the north shall trouble him: therefore he shall go forth with great fury to destroy and utterly to make away many. And **he shall plant the tabernacles of his palace between the seas in the glorious holy mountain**; [Jerusalem in Israel] yet he shall come to his end, and none shall help him.

2 Thessalonians 2:3–4 (bold emphasis and bracketed comments added)

Let no man deceive you by any means: for that day shall not come, except there come a falling away first, and that man of sin be revealed, the son of perdition; Who opposes and exalts himself above all that is called God, or that is worshipped; so that **he as God sits in the temple of God, [which is in Jerusalem] showing himself that he is God**.

The Antichrist will defeat three of the ten nations of the revised Roman Empire, which will reduce the ten nations to seven nations (Daniel 7:23–24, 8:8–9, 8:21–23). The identity of the Antichrist should be glaringly obvious at this time—if there is anyone left who has any knowledge of these prophecies.

1.3. The Second Conquest of the Antichrist

After his initial conquest, the Antichrist will make a seven-year covenant with the nation of Israel during the first three and one-half years of his reign over his newly conquered territory. Then, in the middle of that covenant, he will start another war against the nations that are north and east of him (Daniel 9:27, 11:36–45).

Taking a lesson from recent history, recall what happened with Germany during WWII. The nations surrounding Germany were forced to make a decision to either stand with Germany or fight against it. Italy sided with Germany, while England stood against it.[5] Peaceful nations must always make this difficult decision when faced with imperialistic powers. Just as WWII had its axis and allies, so the future holds a similar fate for the nations in Europe, Asia, and Africa. While the Antichrist will be tightening his grip around his newly conquered territory, the nations to the north and east of him will be preparing for battle, with their correct perception of being next on his list. Germany, Russia, and other nations in Europe not within the revised Roman Empire will be part of this building military alliance (Daniel 11:44; Ezekiel 38–39). Meanwhile the other six nations within the revised Roman Empire will be feeling the pressure, both from the kingdom of the Antichrist, and now the kingdoms of the north and far east, who will be preparing to take a stand against the Antichrist.

Seeing the success of the Antichrist, and also being enamored with him, the six remaining kingdoms inside the revised Roman Empire will decide to give their kingdoms to the Antichrist and make him their military leader (Revelation 17:12–17) in battle against these countries to the north and east (Daniel 11:40–45). Why will they do this? Actually, strangely enough, it will be God who puts it in their hearts to do it (Revelation 17:8, 16–17; 2 Thessalonians 2:11–12). God has a way of using his enemies to destroy each other, and the other enemy he wants to destroy, aside from the Antichrist, in this case is a vast religious system that's been corrupt for a very long time.

Revelation 17:8 (bold emphasis added)

The beast that thou sawest was, and is not; and shall ascend out of the bottomless pit, and go into perdition: **and they that dwell on the earth shall wonder, whose names were not written in the book of life from the foundation of the world, when they behold the beast that was, and is not, and yet is.**

Revelation 17:16–17 (bold emphasis added)

And the ten horns which thou sawest upon the beast, **these shall hate the whore**, and shall make her desolate and naked, and shall eat her flesh, and burn her with fire. For God hath put in their hearts **to fulfill His will**, and to agree, and give their kingdom unto the beast, until the words of God shall be fulfilled.

2 Thessalonians 2:8–12 (bold emphasis added)

And then shall that Wicked be revealed, whom the Lord shall consume with the spirit of his mouth, and shall destroy with the brightness of his coming: Even him, whose coming is after the working of Satan with all power and signs and lying wonders, And with all **deceivableness of unrighteousness in them that perish; because they received not the love of the truth, that they** might be saved. **And for this cause God shall send them strong delusion, that they should believe a lie: That they all might be damned who believed not the truth**, but had pleasure in unrighteousness.

1.4. The Destruction of Mystery Babylon

At the same time all of the political things mentioned above are going on, a religious movement will also be brewing, and this will especially be prevalent in Europe, which currently has a large Roman Catholic base (about 42 percent).[6] The great whore in scripture has been identified by many—even Catholic scholars—as the Roman Catholic Church. (I will elaborate more on this conclusion later.)

The Roman Catholic Church will dominate the Antichrist for a short time during his rise to power over the ten kings, until the middle of Daniel's seventieth week. Then, in the middle of the week, all the kings of the revised Roman Empire will hand over their power to the beast (the Antichrist) because God will put it in their hearts to do so for the specific purpose of destroying Mystery Babylon. In essence the faith in the Roman Catholic Church will crumble from within from the top down. If there are any true believers in Christ within that religious system at that time, they will be forced to make a decision about their church because this religious system will be destroyed so that the Antichrist himself will be worshipped (Revelation 17:16–17). This forced decision is God's extreme way of shoving lukewarm Christians off the fence of complacency, one way, or another! This is why I mentioned earlier that most of the inhabitants in the revised Roman Empire would be enamored with the Antichrist. Since the beast worship is during the last three and one-half years of the reign of the Antichrist, Mystery Babylon must be destroyed at that time (Revelation 13:1–18).

Now what could cause the massive body of the Roman Catholic Church to change its faith? The only thing capable of doing this, in my mind, is if the leaders of the Roman Catholic Church make one of their famous authoritative decisions, as they have done so many times in the past (which again, I will later elaborate on). What kind of decision might they make that would undermine the remnant of truth they have left, and what events could cause them to think so highly of the Antichrist?

At this very point in time, the current pope of the Catholic Church may be holding on to a prophecy known as the Fatima prophecy, which was passed on from one pope to another since 1917.[7] This prophecy was supposed to be opened to the public in 1960, but it was deliberately withheld by the pope who read it back then. The Roman Catholic Church reportedly revealed this prophecy after the failed assassination attempt on Pope John Paul II, but many speculate that it was only revealed in part. All the popes since 1960 have been holding out on something regarding this third secret prophecy given at Fatima in 1917. Some say it states that the pope will betray his flock and turn his sheep over to the slaughter of Lucifer himself.[8] Might this be something any pope would consider withholding?

In any case, when all the kingdoms of the revised Roman Empire unite under the Antichrist, it will be the ultimate demise of the Roman Catholic Church. Something quite incredible is going to convince the people of Europe that the Antichrist is worthy of worship. What on Earth could that be? Perhaps I'm asking the wrong question. What if something *not from Earth* comes into the picture?

The Antichrist will be a mysterious figure indeed. Scripture states that he will be imbued with supernatural power. Concerning the symbol of the beast in Revelation, there are three things represented in the one symbol. First, there is a spirit that comes from the Abyss (Revelation 13:11). Second, there is a mortal human being who is the conduit of the supernatural power of the spirit from the Abyss (Revelation 13:18). Third, the beast as a whole is a reference to both the eighth kingdom of the human Antichrist and the spirit out of the pit (Revelation 17:8–17). So the Antichrist, who will be the leader of the eighth world power, will be empowered by a supernatural being to such a degree that he will be able to make fire come down from the sky, and even cause great miracles to happen (Revelation 13:13–14).

> Revelation 13:11 (bold emphasis added)
>
> And I beheld another **beast coming up out of the earth**; and he had two horns like a lamb, and he spoke as a dragon.
>
> Revelation 13:18 (bold emphasis added)
>
> Here is wisdom. Let him that hath understanding count the number of the beast: for **it is the number of a man**; and his number is Six hundred threescore and six.
>
> Revelation 17:8–17 (bold emphasis and bracketed comments added)
>
> **The beast** that thou sawest was, and is not; and **shall ascend out of the bottomless pit [supernatural spirit]**, and go into perdition: and they that dwell on the earth shall wonder, whose names were not written in the book of life

from the foundation of the world, when they behold the beast that was, and is not, and yet is. And here is the mind which hath wisdom. The seven heads are seven mountains, on which the woman sits. And there are seven kings: five are fallen, and one is, and the other is not yet come; and when he cometh, he must continue a short space. And **the beast that was, and is not, even he is the eighth [eighth kingdom of the human Antichrist]**, and is of the seven, and will go into perdition. And the ten horns which thou sawest are ten kings, which have received no kingdom as yet; but receive power as kings one hour with the beast. These have one mind, and shall give their power and strength unto the beast. These shall make war with the Lamb, and the Lamb shall overcome them: for He is Lord of lords, and King of kings: and they that are with him are called, and chosen, and faithful. And he said unto me, The waters which thou sawest, where the whore sits, are peoples, and multitudes, and nations, and tongues. And the ten horns which thou sawest upon the beast, these shall hate the whore, and shall make her desolate and naked, and shall eat her flesh, and burn her with fire. For God hath put in their hearts to fulfill his will, and to agree, and give their kingdom unto the beast, until the words of God shall be fulfilled.

Revelation 13:12–14 (bold emphasis added)

And he exercises all the power of the first beast before him, and causes the earth and them which dwell therein to worship the first beast, who's deadly wound was healed. **And he doeth great wonders, so that he makes fire come down from heaven on the earth in the sight of men, And deceives them that dwell on the earth by the means of those miracles which he had power to do in the sight of the beast**; saying to them that dwell on the earth, that they should make an image to the beast, which had the wound by a sword, and did live.

Along with the ability to do miracles, the Antichrist will have a strange belief system that doesn't fit with his culture (Daniel 11:37–39). If he's from Lebanon, there's a chance he could be raised either Islamic or Christian (Lebanon consists of a 60 percent Muslim, 39 percent Christian religious demographic),[9] but he clearly won't accept either religion. Instead, he will honor a "strange god."

Daniel 11:37–39 (bold emphasis added)

Neither shall he regard the God of his fathers, nor the desire of women, **nor regard any god: for he shall magnify himself above all. But in his estate shall he honor the God of forces: and a god whom his fathers knew not** shall he honor with gold, and silver, and with precious stones, and pleasant things. Thus shall he do in the most strong holds with a **strange god**, whom

he shall acknowledge and increase with glory: and he shall cause them to rule over many, and shall divide the land for gain.

Whoever this strange god is, it'll be appealing to the Western philosophies of Europe, which is primarily composed of first world countries, all of which have strong scientific ideologies. Just as evolution and Christianity have clashed in the United States, they've clashed in Europe, and the Antichrist will most likely tap into this division with the new religion he will introduce, along with the supernatural authority he will be empowered with to back it up.

The reason I'm spending so much time talking about how the Antichrist will manage to deceive people into believing that he is worthy of worship is that this is where I think the deceptions of alien beings may come into play. This is the encapsulation of this book. The Antichrist will be empowered by a fallen angel (an otherworldly being who has been trapped in a lower dimension of Earth). Reflecting on everything discussed in this book so far, is this not a key indication that the Antichrist will have otherworldly alliances? Consider the fact that one of the specific miracles mentioned of the Antichrist is that he will be able to make fire come down from the sky (Revelation 13:13–14). Does this not denote alliances that span into the heavens? Perhaps the Antichrist will be aligned with beings from other worlds in open contact that will amaze the entire world. Maybe that's why nobody will be able to defeat him—he'll have a technological (as well as supernatural) advantage that no one will be able to overcome.

I'm not alone concerning my opinion that the wars of the future will include otherworldly alliances. People in the highest echelons of government around the world are keenly aware of how such alliances would dramatically affect the outcomes of future wars. Even back in 1955, General Douglas MacArthur voiced his concern about such alliances, and the fact that the people of Earth should stick together, rather than fight among each other.

> "The nations of the world will have to unite, for the next war will be an interplanetary war. The nations of the Earth must some day make a common front against attack by people from other planets."[10]

General Douglas MacArthur was actually prophesying when he stated this, because this statement is exactly what the Antichrist will use for his strategy, to unite the people of Earth against Israel, and against God's invasion to defend Israel (which I will discuss more in depth later). This won't happen until the end of the seven-year reign of the Antichrist, however.

President Ronald Reagan voiced the same concern in an address to the United Nations, given in 1987.

"In our obsession with antagonisms of the moment, we often forget how much unites all the members of humanity. Perhaps we need some outside, universal threat to make us recognize this common bond. I occasionally think how quickly our differences worldwide would vanish if we were facing an alien threat from outside this world."[11]

Also, Mikail Gorbachev, president of the former Soviet Union, had a similar opinion.

"The phenomenon of UFOs does exist, and it must be treated seriously."[12]

Might these national leaders be hinting at something? It appears that they all seem to be fully aware of an external threat, and they're looking at a final confrontation of some sort. The information that is missing, however, is the fact that it's not a simple matter of us (humans) against them (aliens). Alliances will be forged between various organizations (governments, churches, the Jews, and so forth) and beings from other worlds, just as alliances are forged between the nations of Earth nowadays. The result, therefore, will include humans and other-worldly beings on both sides of the confrontation between God and the Antichrist.

Going back to the destruction of the Roman Catholic Church, it's as plain as day to me that an unexpected tangent such as the reality of alien beings from other worlds will be at the heart of the rise of the Antichrist, and the destruction of the Roman Catholic Church. As the greater reality unfolds before the world, the faith of many will crumble, and many will align themselves with the awesome power of what they will see before their very eyes. They will think of the Antichrist as the gateway to the future, a leader established even in the heavens—the person the whole world has been waiting for.

What will people think of the Christians on Earth in those days who refuse this new belief system? They will be thought of as dissidents refusing to accept the truth. Certainly many Christians might be confused about the happenings of the day (at least until they are enlightened by God's faithful angels, and other sources of truth), but the identity of the Antichrist will be clear to any true believer, aliens or no aliens. True Christians will remain faithful to Christ, even if it means death, but the vast majority of the world will be thoroughly amazed with the Antichrist, and I don't think it's just because he'll be charismatic. He'll have a massive deception backing him up all the way to the end—the delusion spoken of in 2 Thessalonians 2:11–12, and this is where it begins. Scripture speaks of three unclean spirits who will be feeding the deception of the Antichrist (Revelation 16:13–16). <u>The primary deception these unclean spirits will be using will have something to do with the new religion that the Antichrist will be advocating, and that new religion will have something to do with beings from other worlds</u>.

Take special note that tied directly in with the turning point of where the Antichrist starts claiming a deitylike nature is a strange event that Daniel connects with an otherworldly battle.

<u>Daniel 8:10–12</u> (bold emphasis and bracketed comments added)

And it [the Antichrist] waxed great, even to the host of heaven [heavenly forces]; and it cast down some of the host and of the stars to the ground, and stamped upon them. Yea, he magnified himself even to the prince of the host, and by him the daily sacrifice was taken away, and the place of his sanctuary was cast down. And an host was given him against the daily sacrifice by reason of transgression, and it cast down the truth to the ground; and it practiced, and prospered.

Here's a possible scenario, as sensationalistic as it may seem. To the amazement of the entire world, extraterrestrial life will become a reality when beings from other worlds introduce themselves in open contact. In their arrival, they will give special attention to the Antichrist. They will bring with them words promising peace, and this might be why Israel agrees to the seven-year peace treaty in the first place. These beings claiming peace will be the bad guys, however, because they will give credence to the Antichrist. (The good angels aren't out of the picture, either. Don't forget my past reference to the activities of God's faithful angels in the end times. The good guys—also beings from other worlds, which people will have to get used to thinking of them as—will be openly communicating with the true believers of the day.)

Shortly after the introduction of these otherworldly beings and after the Antichrist has already consumed three kingdoms from the original ten of the revised Roman Empire, these otherworldly beings will inform the Antichrist of a pending invasion, which they claim will cause great devastation unless the people of Earth cooperate with them. Just as they warned would happen, an invasion ensues—but amazingly, the Antichrist, along with his new allies, will be successful in fending off this otherworldly invasion.

Whether this invasion is from beings allied with God or not is unknown; most likely they will be other fallen beings, and this attack will be a civil war among

Satan's minions. I say this because the Antichrist will successfully defeat this invading force, which would be highly unlikely if he were fighting God's faithful angels.

I know this all seems highly sensational and speculative, but consider the wording of Daniel's prophecy: "He waxed great, even to the host of heaven; and it cast down some of the host and of the stars to the ground, and stamped upon them." This passage is clearly speaking of a battle that originates from the heavens—outer space—and the invaders are cast down to the ground. This event will establish a historical record that will come to play later in the reign of the Antichrist, but initially, the Antichrist will use it to magnify himself. Shortly following this cosmic spectacle, the Antichrist will throw a party in the Jewish temple, putting an end to the daily sacrifices that will be going on there. He'll boldly waltz into the temple and sit in the throne of God (the Mercy Seat of the Ark of the Covenant), and claim that he is God (2 Thessalonians 2:4). Now if that isn't an ego, I don't know what is.

1.5. Mystery Babylon Identified as the Roman Catholic Church

As promised earlier, I said that I'd disclose how I know that the symbol of Mystery Babylon represents the Roman Catholic Church. I don't mean to offend any faithful Catholics, because I'm absolutely certain there are many true Christians who affiliate themselves with Catholicism. But I'm not going to hide what I know about the Roman Catholic Church because I'm afraid of offending anyone.

For starters, the great whore in Revelation 17 is clearly identified as a religious system. It will have a political influence because it will have power over the Antichrist during the first three and one-half years of his reign. Taking just these two basic facts into account, what religion located within the revised Roman Empire could actually exercise political power over governments in that area? Isn't it obvious? The name "Rome" should be a clear hint.

The Roman Catholic religion began in Rome, obviously, but long before Christianity was accepted, Rome had integrated into its governmental system elements of ancient Babylonian Cultism, which actually started with King Nimrod (Genesis 11) and mutated as it passed down from one generation to the next, as well as from one nation to the next. The title of pontiff, and supreme pontiff, which were political offices held in Rome, are actually titles of priesthood in the Babylonian religion. Julius Caesar was made the supreme pontiff of the Etruscan Order in 74 BC, and in 63 BC, he was made supreme pontiff of the Babylonian

Order, thereby becoming heir to the rights and titles of the Attalus, who had made Rome his heir by will.¹³

From this point forward, I'm going to outline a small history lesson I obtained from Finis Dake's *God's Plan for Man*. Finis Dake states the history of Rome so succinctly I'll simply quote directly from the source.

> Thus, the first Roman emperor became the head of the Babylonian priesthood and Rome became the successor of Babylon with Pergamos as the seat of this cult. Henceforth, Rome's religion has been that of Babylon. In the year 218 AD, the Roman army in Syria, having rebelled against Macrinus, elected Elagabalus emperor. This man was a High Priest of the Egyptian branch of Babylonianism. He was shortly afterward chosen Supreme Pontiff by the Romans, and thus the two western branches of the Babylonian apostasy centered in the Roman Emperors who continued to hold this office until 376 AD. At that time, however, the Emperor Gratian for Christian reasons refused it, because he saw that by nature Babylonianism was idolatrous. Thus, religious matters became disorganized until it became necessary to elect someone to fill the office.
>
> Damasus, Bishop of the Christian Church at Rome, was then elected to the office of Supreme Pontiff. He had been bishop for twelve years, having been made such in 366 AD through the influence of the monks of Mount Carmel, a college of the Babylonian religion originally founded by the priests of Jezebel and continued to this day in connection with Rome. Therefore, in 378 AD, the Babylonian system of religion became part of Christendom, for the bishop of Rome, who later became the supreme head of the organized Roman Catholic Church, was also already Supreme Pontiff of the Babylonian Order. All of the teachings of pagan Babylon and Rome were gradually interspersed into the Christian religious organization. Soon after Damasus was made Supreme Pontiff, the rites of Babylon began to come to the front. The worship of the Roman Church became Babylonish, and under him the heathen temples were restored and beautified and the rituals established. Thus, the corrupt religious system under the figure of a woman with a golden cup in her hand, making all the nations drunk with her fornication, is called by God "MYESTERY, BABYLON THE GREAT."¹⁴

Following these words, Finis Dake then proceeds to elaborate on every apostasy introduced into the Roman Catholic Church from this early point in history to the present. These details are no secret to the officials in the Roman Catholic Church that have knowledge of their church history. At the end of this exhaustive list of church doctrines, which had been introduced throughout the past centuries, Dake concludes with the published words of a Roman Catholic Cardinal:

Cardinal Newman, in his book (Page 359) *The Development of the Christian Religion* admits that "Temples, incense, oil lamps, votive offerings, holy water, holidays and seasons of devotions, processions, blessings of fields, sacerdotal vestments, the tonsure (of priests, monks, and nuns), and images are all of pagan origin." Let any honest heart find Scripture for the above practices and see for himself that Romanism is not scriptural. The above chronological list of human inventions disproves the claim of the priests of the Roman Church that their religion was taught by Christ and that the popes have been the faithful custodians of that religion. [15]

So, that all being said, the Roman Catholic Church is clearly the symbol of Mystery Babylon in Revelation, which will come to an end at the hand of the Antichrist.

Does this mean that Catholics aren't saved? Does this mean that Catholics are completely lost? Absolutely not! There have been numerous reforms in the Catholic Church (such as having the Bible in English—the language of the laypeople, for a big start). Knowing a number of Catholics myself, I can certainly attest to the fact that there is wide variation among the beliefs that they profess. Many Catholics are like Mormons—they have no idea what they're affiliating themselves with when they say they're Catholic.

These days the Bible that many Catholics read from is the same Bible that Protestant Christians use (though there was a time in church history when they would've been killed for owning one). In essence the spiritual condition of any person—Catholic, Protestant, or otherwise—can be assessed by the fruit of their lives and the core of their relationship with God. If someone believes in the Bible, and professes and lives out an active, living faith in Christ as mentioned in John 3:16, then that person is saved.

Religion is not what saves people. Christ saves people, and he is able to save anyone provided there is true faith in him, even in spite of peripheral religious dogmatic inaccuracies. It's all a matter of what he finds while he digs around in people's hearts. That's my opinion based on what I see in scripture, anyway.

1.6. Where Will the United States Stand in Relation to the Antichrist?

Going back to the discussions of national alliances in relation to the Antichrist, whom will the United States help in these battles? This is hard to determine. On one side of the coin, the United States has always been allied with England and the majority of Western Europe, which will be inside the revised Roman Empire in the future. Furthermore, if China is still a communist nation at that time, and if it's allied with nations in the north, those nations might be communist nations as well. The United States has always stood against communism. These are two solid reasons why the United States might side with the Antichrist.

On the other side of the coin, the United States has always supported Israel, which the Antichrist will be dead set against (though this opposition won't be noticeable until the middle of his seven-year reign). Second, the United States has never allied itself with other nations based out of fear. During WWII, some nations sided with Germany because they were afraid to oppose Germany, but the United States was not afraid to take a stand against Hitler's tyranny.

It's true that the United States has taken a stand against communism, but it has also stood against imperialism just as much. Germany and Japan are two prime examples. The United States has always stood for freedom, and aligned itself with whatever side represented that freedom best (which has been difficult to determine at times). Judging the imperialistic nature of the Antichrist kingdom, the United States might very well step back from this mess.

Then there's the fact that the United States might have the largest number of new Christians in the world at that time. If anyone will know about the fulfillment of prophecy in those days, people in the United States will be on the forefront, and they certainly won't want to have anything to do with the Antichrist. It'll be obvious who he is. In fact, he'll be slaughtering Christians simply because they're Christians, and I don't see the United States aligning itself with someone like that.

Then again, there are those three lying spirits I mentioned earlier. What effect might they have on the people of the United States?

Concluding the question above, it's impossible to tell who the United States or other nations in the world not within the vicinity of the revised Roman Empire will align with during the course of the seven-year reign of the Antichrist. In Revelation 13:7–8, there is an indication that all the nations of the world will be aligned with the Antichrist, but this isn't the case when one analyzes other scriptures and understands the entire story. First, I'll start with the above-mentioned passage.

Revelation 13:7–8 (bold emphasis added)

And it was given unto him to make war with the saints, and to overcome them: and **power was given him over all kindreds, and tongues, and nations. And all that dwell upon the earth shall worship him, whose**

names are not written in the book of life of the Lamb slain from the foundation of the world.

Initially, this appears to show that all nations will be ruled under the Antichrist, and in a sense, this will be true—but only as it relates to the heart. The deception the Antichrist will be advocating will be global; this is the power and rule that he will be given, but actual governmental authority on the global scale will not be given to him, for three clearly defined reasons:

1. Certain nations (Edom and Moab) will escape his rule (Daniel 11:40–44).

2. At the same time he would supposedly be ruling over every nation of the world, certain countries will be at war with him (Daniel 11:40–44) up until the climax of the battle of Armageddon—which I will disclose in more detail later.

    Daniel 11:40–45 (bold emphasis added)

    And at the time of the end shall **the king of the south push at him**: and the king of the north shall come against him like a whirlwind, with chariots, and with horsemen, and with many ships; and he shall enter into the countries, and shall overflow and pass over. He shall enter also into the glorious land, and many countries shall be overthrown: but **these shall escape out of his hand, even Edom, and Moab, and the chief of the children of Ammon**. He shall stretch forth his hand also upon the countries: and the land of Egypt shall not escape. But he shall have power over the treasures of gold and of silver, and over all the precious things of Egypt: and the Libyans and the Ethiopians shall be at his steps. **But tidings out of the east and out of the north shall trouble him: therefore he shall go forth with great fury to destroy and utterly to make away many**. And he shall plant the tabernacles of his palace between the seas in the glorious holy mountain; yet he shall come to his end, and none shall help him.

3. Multitudes of people from many nations will never take the mark of the beast. Among these, some will never be killed by the Antichrist for not taking this mark. This can be determined by the fact that no one who takes the mark will go into the millennium (Revelation 14:9–11), yet there will be survivors of the tribulation who will see the millennium. These survivors must therefore be those who never took the mark of the beast (Zechariah 14:16–21; Joel 2:29–32; Isaiah 2:1–4; Matthew 25:31–46; and other scriptures).

Revelation 14:9–11 (bold emphasis added)

And the third angel followed them, saying with a loud voice, **If any man worship the beast and his image, and receive his mark in his forehead, or in his hand, the same shall drink of the wine of the wrath of God, which is poured out without mixture into the cup of his indignation; and he shall be tormented with fire and brimstone in the presence of the holy angels, and in the presence of the Lamb: And the smoke of their torment will ascend up for ever and ever**: and they will have no rest day nor night, who worship the beast and his image, and whosoever received the mark of his name.

Zechariah 14:16–21

And it shall come to pass, that **every one that is left of all the nations which came against Jerusalem shall even go up from year to year to worship the King, the LORD of hosts, and to keep the feast of tabernacles**. And it shall be that whoso will not come up of all the families of the earth unto Jerusalem to worship the King, the LORD of hosts, even upon them shall be no rain. And if the family of Egypt go not up, and come not, that have no rain; there shall be the plague, tabernacles. This shall be the punishment of Egypt, and the punishment of all nations that come not up to keep the feast of tabernacles. In that day shall there be upon the bells of the horses, HOLINESS UNTO THE LORD; and the pots in the LORD's house shall be like the bowls before the altar. Yea, every pot in Jerusalem and in Judah shall be holiness unto the LORD of hosts: and all they that sacrifice shall come and take of them, and seethe therein: and in that day there shall be no more the Canaanite in the house of the LORD of hosts.

Isaiah 2:1–4 (bold emphasis added)

The word that Isaiah the son of Amoz saw concerning Judah and Jerusalem. And it shall come to pass in the last days, that the mountain of the LORD's house shall be established in the top of the mountains, and shall be exalted above the hills; and all nations shall flow unto it. **And many people shall go and say, Come ye, and let us go up to the mountain of the LORD, to the house of the God of Jacob; and he will teach us of his ways, and we will walk in his paths**: for out of Zion shall go forth the law, and the word of the LORD from Jerusalem. And he shall judge among the nations, and shall rebuke many people: and they shall beat their swords into plowshares, and their spears into pruning hooks: nation shall not lift up sword against nation, neither shall they learn war any more.

Regardless of whom North America, South America, Australia, or other nations decide to align with, the Antichrist will succeed in defeating the nations to the north and east of him during the last three and one-half years of his reign. After these victories, he will be known as the chief prince of Meshech and Tubal (Ezekiel 38–39). All of the newly conquered territory he will acquire will form the eighth world power, as in John's visions—a global force unlike any in the history of the world. However, will it encompass the whole world? No.

1.7. The Antichrist's Frustration with Israel

Israel will be a pain in the side of the Antichrist. He will start out making a seven-year covenant of peace with Israel, which will be devastating for Israel because it will be by this peace that he will destroy many of them (Daniel 8:25, 9:27; Matthew 24:15–26). The Antichrist won't wipe them out by a long shot, though, because God will preserve his "remnant," as God lovingly calls them.

Daniel 11:41 states that ancient Edom and Moab (modern-day Jordan and Arabia) will escape the Antichrist (Daniel 11:41). Delving further, in Isaiah 16:1–5, Ezekiel 20:33–35, Matthew 24:15–21, and Revelation 12:5–17, it is stated that Israel will flee for protection into the wilderness (Edom and Moab—places that escape the Antichrist) which interestingly are currently Arab nations, obviously—duh, Arabia. Arabs, therefore, will be helping the Jews during this troublesome time. It takes the Antichrist in their face to make this happen, but hey, they'll finally get along! I think the crux of what will make this possible is that there will be many more Christian Arabs in the future, as well as more Messianic Jews (Christians), and they will find their common ground in Christ (even as they currently do, but on a much smaller scale). Way cool! Currently the few Arab Christians there are in this region are commonly anti-Semitic (not all, but many of them), making their flavor of Christianity highly tarnished.[16] (The first time I heard about anti-Semitic Christians, I did a double-take; think about it—you'd hate the very God you worship, because Jesus is a Jew!) Apparently God will have to do something about this inconsistency in the faith of anti-Semitic Arabic Christians, in order to get them to help the Jews in the future.

> Daniel 11:41 (bold emphasis added)
>
> He shall enter also into the glorious land, and many countries shall be overthrown: but **these shall escape out of his hand, even Edom, and Moab**, and the chief of the children of Ammon.

Isaiah 16:1–5 (bold emphasis added)

Send ye the lamb to the ruler of the land from Sela to the wilderness, unto the mount of the daughter of Zion. For it shall be that, as a wandering bird cast out of the nest, so the daughters of Moab shall be at the fords of Arnon. Take counsel, execute judgment; make thy shadow as the night in the midst of the noonday; **hide the outcasts; betray not him that wanders. Let mine outcasts dwell with thee, Moab; be thou a covert to them from the face of the spoiler**: for the extortioner is at an end, the spoiler ceases, the oppressors are consumed out of the land. And in mercy shall the throne be established: and he shall sit upon it in truth in the tabernacle of David, judging, and seeking judgment, and hasting righteousness.

Ezekiel 20:33–35 (bold emphasis added)

As I live, says the Lord GOD, surely with a mighty hand, and with a stretched out arm, and with fury poured out, will I rule over you: And I will bring you out from the people, and will gather you out of the countries wherein ye are scattered, with a mighty hand, and with a stretched out arm, and with fury poured out. And **I will bring you into the wilderness of the people, and there will I plead with you face to face.**

Revelation 12:5–17 (bold emphasis added)

And she brought forth a man child, who was to rule all nations with a rod of iron: and her child was caught up unto God, and to his throne. **And the woman fled into the wilderness, where she hath a place prepared of God that they should feed her there a thousand two hundred and threescore days**. And there was war in heaven: Michael and his angels fought against the dragon; and the dragon fought and his angels, and prevailed not; neither was their place found any more in heaven. And the great dragon was cast out, that old serpent, called the Devil, and Satan, which deceives the whole world: he was cast out into the earth, and his angels were cast out with him. And I heard a loud voice saying in heaven, Now is come salvation, and strength, and the kingdom of our God, and the power of his Christ: for the accuser of our brethren is cast down, which accused them before our God day and night. And they overcame him by the blood of the Lamb and by the word of their testimony; and they loved not their lives unto the death. Therefore rejoice ye heavens, and ye that dwell in them. Woe to those who inhabit the earth and of the sea! For the devil is come down unto you, having great wrath, because he knows that he hath but a short time. And when the dragon saw that he was cast unto the earth, he persecuted the woman which brought forth the man child. **And to the woman were given two wings of a great eagle, that she might fly into the wilderness, into her place, where she is nourished for a time, and times, and half a time, from the face of the serpent**. And the serpent cast out of his mouth water as a flood after the woman, that he might cause her to be carried away of the flood. And the earth helped the woman, and the earth opened her mouth, and swallowed up the flood which the dragon cast out of his mouth. And the dragon was wroth with the woman, and went to make war with the remnant of her seed, which keep the commandments of God, and have the testimony of Jesus Christ.

The Bible even states the exact location of where the Jews will flee. Sela, the ancient capital of Edom, is now called Petra; it's an abandoned desert fortress carved out of solid rock (amazingly surrealistic too, I might add). This ancient city nestled into the rocks of the desert canyons will be the haven of the Jews during the great tribulation (Isaiah 16:1–5, and referred to in Isaiah 26:20–21, 63:1–8).

Isaiah 16:1 (bold emphasis added)

Send ye the lamb to the ruler of the land from Sela to the wilderness, unto the mount of the daughter of Zion. For it shall be that, as a wandering bird cast out of the nest, so **the daughters of Moab shall be at the fords of Arnon.**

The funny thing is, the Antichrist will probably know about all of these prophecies, being familiar with scripture as I'm sure he'll be, but he won't be able to do anything about them. He'll try to stop the Jews during their escape by sending a military force after them, but God will cause a massive earthquake to open up the ground and swallow them up (Revelation 12:13–16).

Revelation 12:13–17 (bold emphasis added)

And when the dragon saw that he was cast unto the earth, he persecuted the woman which brought forth the man child. And to the woman were given two wings of a great eagle, that she might fly into the wilderness, into her place, where she is nourished for a time, and times, and half a time, from the face of the serpent. **And the serpent cast out of his mouth water as a flood after the woman, that he might cause her to be carried away of the flood. And the earth helped the woman, and the earth opened her mouth, and swallowed up the flood which the dragon cast out of his mouth.** And the dragon was wroth with the woman, and went to make war with the remnant of her seed, which keep the commandments of God, and have the testimony of Jesus Christ.

Then there's the bolder Jews, who decide to stand their ground in Israel, right where the Antichrist will have his capital established during the second half of his reign. Even they will be protected from the Antichrist, despite the fact that they'll be right there under his nose. They will be protected by both the two witnesses (Enoch and Elijah will return to Earth in those days to preach the Gospel and protect the Jews—Revelation 11:3–13), and the wars in the north and east will deter the Antichrist from focusing his attacks on them until the very tail end of the great tribulation.

> Revelation 11:3–13 (bold emphasis added)
>
> **And I will give power unto my two witnesses**, and they shall prophesy a thousand two hundred and threescore days, clothed in sackcloth. These are the two olive trees, and the two candlesticks standing before the God of the earth. **And if any man will hurt them, fire will proceed out of their mouth, and devour their enemies: and if any man will hurt them, he must in this manner be killed. These have power to shut heaven that it rain not in the days of their prophecy: and have power over waters to turn them to blood, and to smite the earth with all plagues, as often as they will**. And when they shall have finished their testimony, the beast that will ascend out of the bottomless pit shall make war against them, and shall overcome them, and kill them. And their dead bodies shall lie in the street of the great city, which spiritually is called Sodom and Egypt, where also our Lord was crucified. And they of the people and kindreds and tongues and nations shall see their dead bodies three days and an half, and shall not suffer their dead bodies to be put in graves. And they that dwell upon the earth shall rejoice over them, and make merry, and shall send gifts one to another; because these two prophets tormented them that dwelt on the earth. And after three days and an half the Spirit of life from God entered into them, and they stood upon their feet; and great fear fell upon them which saw them. And they heard a great voice from heaven saying unto them, Come up hither. And they ascended up to heaven in a cloud; and their enemies beheld them. And the same hour was there a great earthquake, and the tenth part of the city fell, and in the earthquake were slain of men seven thousand: **and the remnant were affrighted, and gave glory to the God of heaven.**

Certainly all of this will irk the Antichrist to no end. He'll be successful in waging war and destroying people all over the world, but the Jewish people, whom he will hate more than anyone else, will be right in his own backyard, living in the same city where he will have his capital established, and he still won't be able to destroy them.

1.8. The Technology of the Antichrist Kingdom

I mentioned before that the Antichrist might have a technological advantage in his battles because of his alliance with technologically advanced beings from other worlds. But I didn't disclose the full depth of everything the Bible says about the technology utilized in the Antichrist kingdom. When the Antichrist assumes control over his new kingdom, one of the first mandates given by the beast from the Abyss (the supernatural being helping the Antichrist) will be to create an image of the beast (a representation of the Antichrist and his kingdom). This man-made thing will be intelligent, able to speak, and serve the Antichrist by exerting administrative control over his kingdom. He will regulate the economic system of his kingdom by forcing all his subjects to accept a special identification mark (the mark of the beast). Those without the mark (or the number or name of the beast) will not be allowed to buy or sell goods within his kingdom.

> Revelation 13:14–16 (bold emphasis added)
>
> And deceives them that dwell on the earth by the means of those miracles which he had power to do in the sight of the beast; saying to them that dwell on the earth, **that they should make an image to the beast**, which had the wound by a sword, and did live. And he had power to **give life unto the image of the beast**, that **the image** of the beast **should both speak, and cause that as many as would not worship the image of the beast should be killed**. And **he causes all**, both small and great, rich and poor, free and bond, **to receive a mark** in their right hand, or in their foreheads: And **that no man might buy or sell, save he that had the mark, or the name of the beast, or the number of his name.**

This man-made, intelligent being will be an active participant in the administration of the Antichrist kingdom, but it won't actually be alive because it's never mentioned as being destroyed by God along with the beast (the Antichrist) and the false prophet.

> Revelation 19:20
>
> And the beast was taken, and with him the false prophet that wrought miracles before him, with which he deceived them that had received the mark of the beast, and them that worshipped his image. These both were cast alive into a lake of fire burning with brimstone.

I'm a fairly savvy computer guy, but I don't think it takes a computer person to see that this image of the beast will be a technological wonder. It's a man-made, intelligent being—the exact description of a computer given by someone who had no idea what a computer was. These days just about everyone familiar with prophecy, even a little bit, thinks the mark of the beast will be some sort of computer chip. I'd put a reference for this statement, but anyone with Internet access can simply search the Internet for "mark of the beast" and see that many of the links will give references to computer chips. The UPC (United Purchase Seal) barcode will also pop up; today's standard for barcode scanners. The UPC apparently has "666" encoded into it for some technical reason, which is an interesting coincidence, since the UPC has to do with buying and selling.

[margin note: 666 encoded in all bar codes!]

Even in my job in the military, I happen to know that PKI (Public Key Infrastructure) technology is headed in the direction of an embedded computer chip. My personal military ID card, which is now called the "Common Access Card," (CAC) contains a gold-colored computer chip on the front of it. The CAC is used for logging into military networks, accessing PKI encrypted Web sites, and reading PKI encrypted e-mail. I personally doubt the embedded-chip technology will ever be implemented in the United States, but it looks as if something like this will be exactly what the Antichrist kingdom will utilize for identification and economic administration. While most people understand the mark of the beast as a form of technology, I'm the only person that I know of who thinks that the image of the beast will be an artificially intelligent android.

Of course the image of the beast being an artificially intelligent android isn't particularly significant in itself, other than the fact that I'm predicting what the image of the beast will be, and this will add to the credibility of the interpretation of the other information I've highlighted in this book if it comes to pass.

1.9. Could the Antichrist be a Modern-Day Nephilim?

For many years, there have been numerous speculations about the Antichrist. When will he come to power? Where will he come from? Will he rule from Rome, or some country in Europe, or even the United States? Will he be one of the popes or a president of the United States? Will he be a mortal human being, or something else, such as the reincarnation of Judas Iscariot, or the physical incarnation of Satan, or perhaps even a being from the underworld—from the bottomless pit? Might he be a modern-day Nephilim?

Concerning the first three questions, I've already stated that the Antichrist won't rise to power until after the rapture of the church and the formation of the ten nations of the revised Roman Empire. As for the location of his initial ascent into world dominance, the Antichrist will make his entrance in the Syrian division of the ancient Grecian Empire, which today is either Syria or Lebanon. Because his location is clearly determined with such precision, it's obvious that he won't be a pope, or a president of the United States, nor will he ever establish his capital in Rome, or in the United States, or anywhere other than where scripture clearly states. The city of Babylon will be rebuilt and established as the capital of the Antichrist kingdom during the first three and one-half years of Daniel's seventieth week (Isaiah 13:6, 13:9–13, 14:1–11; Zechariah 5:5–11; Revelation 16:17–21; 18:1–24). During the second three and one-half years of his reign, the Antichrist will relocate his capital to Jerusalem (Daniel 9:27, 11:40–45; 2 Thessalonians 2:3–4; Revelation 13).

As for the humanity of the Antichrist, he will definitely be a mortal human, though he won't be *humane*. Revelation 13:18 states that his number (666) is the number of a man. He will also be slain by Christ when Christ returns at his second advent (Daniel 7:11; Isaiah 11:4; 2 Thessalonians 2:7–8), which wouldn't be possible to do if he were some completely immortal angelic being.

In reference to the idea that the Antichrist will be a reincarnation of Judas Iscariot, Hebrews 9:27 states that it is appointed unto men only once to die, and after that comes the judgment—so Judas Iscariot is out of the picture, and so is anyone else who could be reincarnated, for that matter.

Does the Nephilim theory have any substance to it? Might the Antichrist be a modern-day Nephilim? After weighing everything out, I think this might actually be a possibility, because the Nephilim (mortal, half-breed human-angel hybrids first mentioned in Genesis 6) are actually called "mighty men" and "men of renown" in scripture (Genesis 6:4), so they are considered to be men, even though they're genetically diluted.

Scripture also states that the final conflict between God and Satan will be between the seed of the woman, Jesus, and the seed of Satan, the Antichrist (Genesis 3:15). The seed of the woman was a real, physical, biological seed, so why would the other seed *not* be a real, physical, biological seed, since all the elements necessary to make that possible are available?

> Genesis 6:4 (bold emphasis added)
>
> There were giants in the earth in those days; and also after that, when the sons of God came in unto the daughters of men, and they bare children to them, the same became **mighty men** which were of old, **men of renown**.

Genesis 3:15 (bold emphasis and bracketed comments added)

And I will put enmity between thee and the woman, and **between thy [Satan's] seed** and **her [Eve's] seed**; it shall bruise thy head, and thou shalt bruise his heel.

Since the Nephilim were actually the biological offspring of fallen angels, it stands as a possibility that Satan has been saving his personal seed for a long time—waiting for the right time to play this last trump card. He clearly wasn't participating with the rebel angels who committed the sin of fornication with humans documented in Genesis 6 (an especially reprehensible sin, apparently, because those fallen angels who were committing it were apprehended and chained up, while other fallen angels are still allowed to roam free). Certainly he had it within him to commit this sin, but he specifically didn't, and I suspect this is the reason. If Satan committed the special sin that the other angels committed in Genesis 6, he would've been chained up in hell with the rest of them long ago, but he wasn't (Jude 1:6; 2 Peter 2:4).

Jude 1:6 (bold emphasis added)

And **the angels who kept not their first estate, but left their own habitation, he hath reserved in everlasting chains under darkness** unto the judgment of the great day.

2 Peter 2:4 (bold emphasis added)

For if **God spared not the angels that sinned, but cast them down to hell, and delivered them into chains of darkness**, to be reserved unto judgment...

So perhaps the Antichrist really will be the biological son of Satan. This can actually explain a lot, because it gives the Antichrist a unique affiliation with the heavenly realm. It explains his strange religion and qualifies his special place as its appointed leader, a sort of messiah for the New Age. It also explains even more about the origin of his supernatural power from Satan (2 Thessalonians 2:7–12; Daniel 11:35–39; Revelation 13:1–4). Statements that otherworldly visitors may make about him could actually be true statements, even though deception will be behind everything they say. He will be special to them because he will actually be from them—a chosen one, so to speak—as they might very well state and even be able to prove.

With all the facts taken into consideration, the Antichrist bears the exact resemblance of a Nephilim. He could literally be the son of perdition (2 Thessalonians 2:1–12), being just as evil as the Nephilim were (Genesis 6:5), certainly a mighty man of renown, with his fierce countenance (Daniel 8:23) and peculiar command of supernatural power—all the stuff of Greek legends, which were inspired by the original Nephilim. He will even come from the same ancient Grecian Empire where all those satanic Nephilim once flourished, and the main thing he'll be trying to do during his entire reign will be exactly the same thing the Nephilim did—fill the world with violence in a mad torrent of conquest, demanding nothing less than worship as a god.

On this particular topic, I disagree with Finis Dake because he was rather adamant in his view that the Antichrist will be a normal, mortal man. I'm not sure why he wasn't willing to see the possibility of the Antichrist being a Nephilim, since he understood what the Nephilim really were. I agree with Dake that the Antichrist will be a mortal man, but I can't deny the fact that he is certainly not normal, and that he bears the earmark of a Nephilim in every way. Just because the characteristics and identity of the Antichrist can be understood with a purely natural interpretation of scripture, and Satan is fully capable of exercising his power through someone without having a biological son, doesn't mean he won't actually have a biological son. It's possible, and scripture seems to indicate this, in my opinion.

Perhaps this is what lies behind the alien-abduction phenomenon and the genetic experimentation at the core of it all. Perhaps the fallen sons of God, Satan's cronies, are ironing out all the Nephilim bugs of the past (getting the size ratio corrected, ensuring there will be five fingers and toes instead of six—2 Samuel 21:20 and others). As I stated in *Aliens in the Bible*, the alien abduction phenomenon is probably not some bizarre pastime of ultraintelligent beings conducting pranks relative to cow tipping in Wisconsin.

2. The Final Conquest of the Antichrist, and Second Advent of Christ

Near the end of the second three and one-half years of the reign of the Antichrist, after the Antichrist absorbs the nations to the north and east of him into his own kingdom, he'll finally focus his attention back on the Jews again. Up until this point, the Jews will be protected either by the two witnesses or by the Arabs in the nations that will thus far escape the Antichrist, but now their good fortune will appear to change.

The two witnesses will be wreaking havoc in Israel for the second three and one-half years, while the Antichrist will be busy fighting to the north and east, but the two witnesses will finally be overcome by the beast that ascends from the bottomless pit. This will occur at the same time the Antichrist claims victory over the north and east (Revelation 11:3–10). The two witnesses will be murdered, and their bodies will lie in a street in Jerusalem for three days while the majority of the world celebrates their deaths.

> Revelation 11:7–10
>
> And when they shall have finished their testimony, the beast that will ascend out of the bottomless pit shall make war against them, and shall overcome them, and kill them. And their dead bodies shall lie in the street of the great city, which spiritually is called Sodom and Egypt, where also our Lord was crucified. And they of the people and kindreds and tongues and nations shall see their dead bodies three days and an half, and shall not suffer their dead bodies to be put in graves. And they that dwell upon the earth shall rejoice over them, and make merry, and shall send gifts one to another; because these two prophets tormented them that dwelt on the earth.

Soon after the big celebration is over, the Antichrist will prepare to mount a full-scale attack against the Jews, but right at the same time that he starts to mount this attack, he may receive word from his otherworldly allies of another impending invasion from the heavens. This is the second invasion I mentioned earlier, which will be much larger than the first one. This invasion will be directly linked with the attack about to be waged on the Jews (Jesus will be returning with the armies of heaven to rescue them), though the rest of the world might not be aware of this. All they will know, if anything, is that Earth will be invaded by a hostile force.

I always wondered exactly how the Antichrist would convince the entire world—every nation on the planet—to align itself under his military command in the battle of Armageddon against Israel and the armies of heaven. What will this look like? What will have to unfold to make this scenario possible? Certainly the truth of the identity of the approaching otherworldly armada would have to be obscured in some manner in order to convince everyone to unite and fight God and the armies of heaven.

As I already stated, I believe the Antichrist will have an accomplished precedent of being victorious in battle even over otherworldly opponents, as described in Daniel 8:10–12. In addition, the fact that the Antichrist and much of his army will still be alive at the end of the second three and one-half years gives an indication that they may also be able to fight successfully against the creatures of the Abyss that will be unleashed. The Antichrist will capitalize on these precedents, using them to convince the world that the attack that is coming is a satanic attack bent on the destruction of the world. Many people will actually believe that the Antichrist is the messiah with his incredible charisma, seemingly invincible leadership and strength, and most of all, his unique heavenly origins and allies, along with his supernatural power. Nothing in those days will be as it seems.

The three unclean spirits will have completed their work, successfully deceiving everyone in the world who rejects the truth of who God really is (Revelation 16:13–16). Thankfully God's servants will have done their work as well! They will be empowered with miraculous gifts to do even greater miracles than Jesus did (John 14:12). With their miraculous powers, they will be spreading the Gospel in an unprecedented revival during the great tribulation that will save more people on Earth than in any revival in the history of the world (Joel 2:29–32; Acts 2:17–18; John 14:12).

John 14:12

Verily, verily, I say unto you, he that believeth on me, the works that I do shall he do also; and **greater works than these shall he do**; because I go unto my Father.

Joel 2:29–32

And also upon the servants and upon the handmaids in those days will I pour out my spirit. And I will show wonders in the heavens and in the earth, blood, and fire, and pillars of smoke. The sun shall be turned into darkness, and the moon into blood, before the great and the terrible day of the LORD come. And it shall come to pass, that whosoever shall call on the name of the LORD shall be delivered: for in Mount Zion and in Jerusalem shall be deliverance, as the LORD hath said, and in the remnant whom the LORD shall call.

Acts 2:17–21

And it shall come to pass in the last days, says God, I will pour out of my Spirit upon all flesh: and your sons and your daughters shall prophesy, and your young men shall see visions, and your old men shall dream dreams: And on my servants and on my handmaidens I will pour out in those days of my Spirit; and they shall prophesy: And I will show wonders in heaven above, and signs in the earth beneath; blood, and fire, and vapor of smoke: the sun shall be turned into darkness, and the moon into blood, before that great and notable day of the Lord come: and it shall come to pass, that whosoever shall call on the name of the Lord shall be saved.

Concerning the impending invasion, the Antichrist will successfully deceive many people on Earth into believing that he will be their savior who will save them from an evil attack about to take place (Ezekiel 38–39; Zechariah 14; Joel 3; Revelation 14:9–10, 19:11–21). But of course, when Jesus returns, the Antichrist and the false prophet, as well as all their followers who accepted the mark of the beast, will be completely destroyed (Zechariah 14:1–5; Jude 14; Revelation 19:11–21).

Revelation 14:9–10

And the third angel followed them, saying with a loud voice, If any man worship the beast and his image, and receive his mark in his forehead, or in his hand, The same shall drink of the wine of the wrath of God, which is poured out without mixture into the cup of his indignation; and he shall be tormented with fire and brimstone in the presence of the holy angels, and in the presence of the Lamb.

There won't even be a contest in the battle of Armageddon—when Jesus returns, he will put an end to the madness. This well-famed battle spoken of in heated circles for millennia will result in more bloodshed than any other battle ever waged on Earth. The entire valley floor will be filled with a lake of blood about 200 miles long, and as deep as a horse's bridle—which is about five feet deep (Revelation 14:20). Following this battle, birds will gorge themselves on the flesh of corpses for a considerable amount of time (Revelation 19:18).

Revelation 14:20

And the winepress was trodden without the city, and blood came out of the winepress, even unto the horse bridles, by the space of a thousand and six hundred furlongs.

Revelation 19:18

That ye may eat the flesh of kings, and the flesh of captains, and the flesh of mighty men, and the flesh of horses, and of them that sit on them, and the flesh of all men, both free and bond, both small and great.

I find it interesting to note that people who aren't Christians will frequently be completely turned off to the name of Jesus, but mystical terminology referring to prophecies about the Antichrist or the battle of Armageddon will perk their ears up with interest. Movies such as *Prophecy, End of Days, The Seventh Sign*, and so forth are works of fiction, yet they draw considerable interest in the secular community. Isn't the truth interesting enough—with the Nephilim Antichrist, his otherworldly allies, and battles with strange creatures from other dimensions within Earth? Imagine the climax of such a movie, building up to the battle of Armageddon, with all the nations in the entire world united against an otherworldly invasion they are led to believe is an evil army. Then, in the last moment, they suddenly come to the horrific conclusion that it is God, his mighty angels, and his saints of the past that they will be fighting! "Oops, I guess I'm on the wrong side!" Some will realize all too late. It will be like *War of the Worlds*, but the people on Earth in those days will be the bad guys—now that's a twist. Wake up, Hollywood—truth is stranger than fiction!

The real story is far more captivating than any work of fiction, and the ending is perhaps the most difficult to imagine, with its surreal beauty and awe. Jesus, the man of peace, as well as the ultimate warrior with unlimited power, will return to Earth in all his brilliant glory and splendor. He will lead the armies of heaven—riding white horses across the sky, most likely amid a myriad of otherworldly beings and various spacecraft, all fighting each other. Lasers, bolts of electricity, plasma cannons, photon torpedoes, and who knows what else will be blasting away in all directions. The battles in the heavens that have been going on for thousands of years will finally become visible to the people of Earth, and all hell will literally break loose.

Does the story end here? No, quite the contrary—only now does the story begin.

PART IV
THE HEAVENS OF THE FUTURE

Currently most people are blind in their perception of the greater reality of the heavens. A barrier exists, which is both spatially (the vast distances in the cosmos) and dimensionally defined; it serves as a protective barrier for humanity (keeping us in, and others out, until the appointed time). God uses this veil for protection, but he also uses it for another reason as well: to test faith.

There's an extra blessing for those who learn to live according to the unseen realms (Hebrews 11; John 20:29).

> Hebrews 11:1–10
>
> Now faith is the substance of things hoped for, the evidence of things not seen. For by it the elders obtained a good report. Through faith we understand that the worlds were framed by the word of God, so that things which are seen were not made of things which do appear. By faith Abel offered unto God a more excellent sacrifice than Cain, by which he obtained witness that he was righteous, God testifying of his gifts: and by it he being dead yet spoke. By faith Enoch was translated that he should not see death; and was not found, because God had translated him: for before his translation he had this testimony, that he pleased God. But without faith it is impossible to please him: for he that cometh to God must believe that he is, and that he is a rewarder of them that diligently seek him. By faith Noah, being warned of God of things not seen as yet, moved with fear, prepared an ark to the saving of his house; by which he condemned the world, and became heir of the righteousness which is by faith. By faith Abraham, when he was called to go out into a place which he should after receive for an inheritance, obeyed; and he went out, not knowing whither he went. By faith he sojourned in the land of promise, as in a strange country, dwelling in tabernacles with Isaac and Jacob, the heirs with him of the same promise: for he looked for a city which hath foundations, whose builder and maker is God.
>
> John 20:29 (bold emphasis added)
>
> Jesus said unto him, Thomas, because thou hast seen me, thou hast believed: **blessed are they that have not seen, and yet have believed.**

If God were clearly visible to the world, many people might claim allegiance to him based on the facts (as will be the case in the future—Isaiah 2:3; Micah 4:2), but it would be a belief derived through the natural senses, rather than through the hope that there really is a God who loves and cares about us. Those who demonstrate faith without having an absolute knowledge of the facts are more worthy of their reward.

Isaiah 2:3

And many people shall go and say, Come ye, and let us go up to the mountain of the LORD, to the house of the God of Jacob; and he will teach us of his ways, and we will walk in his paths: for out of Zion shall go forth the law, and the word of the LORD from Jerusalem.

Micah 4:2

And many nations shall come, and say, Come, and let us go up to the mountain of the LORD, and to the house of the God of Jacob; and he will teach us of his ways, and we will walk in his paths: for the law shall go forth of Zion, and the word of the LORD from Jerusalem.

God wants to know what we hope for without letting us know the concrete facts about his existence. Scripture clearly states that God wants us to seek him out (Acts 17:24–28), and there would be no seeking if we already knew for a fact that he existed.

Acts 17:24–28 (bold emphasis added)

God that made the world and all things therein, seeing that he is Lord of heaven and earth, dwelleth not in temples made with hands; Neither is worshipped with men's hands, as though he needed any thing, seeing he giveth to all life, and breath, and all things; And hath made of one blood all nations of men for to dwell on all the face of the earth, **and hath determined** the times before appointed, and the bounds of their habitation; **That they should seek the Lord, if haply they might feel after him, and find him,** though he be not far from every one of us: For in him we live, and move, and have our being; as certain also of your own poets have said, For we are also his offspring.

If we hope for a God of selfish desires, of hatred and pride—if we hope for a push-button God who says yes to all our prayers without question, or a God who doesn't care what we do, then he's not interested in revealing himself to us because we won't be interested to find him for who he truly is. It's all a matter of what's in our hearts (Psalms 44:21, 64:6, 139:23–24; Jeremiah 17:10), which determines what we hope for. So by keeping humanity in the dark, and giving the truth only to those who seek after it with a hope that God is good (Jeremiah 29:13; Matthew 7:7; Luke 11:9), the condition and worthiness of the human heart becomes weightily evident.

Psalms 44:21

Shall not God search this out? For he knows the secrets of the heart.

Psalms 64:6

They search out iniquities; they accomplish a diligent search: both the inward thought of every one of them, and the heart, is deep.

Psalms 139:23–24

Search me, O God, and know my heart: try me, and know my thoughts: And see if there be any wicked way in me, and lead me in the way everlasting.

Jeremiah 17:10

I the LORD search the heart, I try the reins, even to give every man according to his ways, and according to the fruit of his doings.

Jeremiah 29:13

And ye shall seek me, and find me, when ye shall search for me with all your heart.

Matthew 7:7–8

Ask, and it shall be given you; seek, and ye shall find; knock, and it shall be opened unto you: For every one that asks receives; and he that seeks finds; and to him that knocks it shall be opened.

Luke 11:9–10

And I say unto you, ask, and it shall be given you; seek, and ye shall find; knock, and it shall be opened unto you. For every one that asks receives; and he that seeks finds; and to him that knocks it shall be opened.

It might seem unfair that some people are given more light (revelation of God) than others are, but God ensures fairness in two ways. First, he gives an extra blessing to those who exercise faith in someone they can't see (John 20:29). Second, he will judge us only according to the light that we have received. If we respond to the light that God gives us to start with, he will give us more light (Matthew 25:14–29). Therefore anyone who truly wants to know if there is a God and earnestly seeks, responding to the light that he or she has, will receive more light. Those who lose interest in the search, for whatever reason, shall have the small amount of light that they have received taken away from them. This is explained in the Bible where it states that Satan is a thief of truth, and God will allow him to steal the little bit of truth we have when we refuse to respond to the little bit of light God gives us to start with (Matthew 13:3–8, 19–23).

Matthew 25:14–29*

For the kingdom of heaven is as a man travelling into a far country, who called his own servants, and delivered unto them his goods. And unto one he gave five talents, to another two, and to another one; to every man according to his several ability; and straightway took his journey. Then he that had received the five talents went and traded with the same, and made them other five talents. And likewise he that had received two, he also gained other two. But he that had received one went and digged in the earth, and hid his lord's money. After a long time the lord of those servants cometh, and reckoneth with them. And so he that had received five talents came and brought other five talents, saying, Lord, thou deliveredst unto me five talents: behold, I have gained beside them five talents more. His lord said unto him, Well done, thou good and faithful servant: thou hast been faithful over a few things, I will make thee ruler over many things: enter thou into the joy of thy lord. He also that had received two talents came and said, Lord, thou deliveredst unto me two talents: behold, I have gained two other talents beside them. His lord said unto him, Well done, good and faithful servant; thou hast been faithful over a few things, I will make thee ruler over many things: enter thou into the joy of thy lord. Then he which had received the one talent came and said, Lord, I knew thee that thou art an hard man, reaping where thou hast not sown, and gathering where thou hast not strawed: And I was afraid, and went and hid thy talent in the earth: lo, there thou hast that is thine. His lord answered and said unto him, Thou wicked and slothful servant, thou knewest that I reap where I sowed not, and gather where I have not strawed: Thou oughtest therefore to have put my money to the exchangers, and then at my coming I should have received mine own with usury. Take therefore the talent from him, and give it unto him which hath ten talents. For unto every one that hath shall be given, and he shall have abundance: but from him that hath not shall be taken away even that which he hath.

The talents may be synonymous with either blessings that God bestows upon us or truths that he imparts to us.

Matthew 13:3–8

And he spake many things unto them in parables, saying, Behold, a sower went forth to sow; And when he sowed, some seeds fell by the way side, and the fowls came and devoured them up: Some fell upon stony places, where they had not much earth: and forthwith they sprung up, because they had no deepness of earth: And when the sun was up, they were scorched; and because they had no root, they withered away. And some fell among thorns; and the thorns sprung up, and choked them: But other fell into good ground, and brought forth fruit, some an hundredfold, some sixtyfold, some thirtyfold. Who hath ears to hear, let him hear.

Matthew 13:19–23

When any one heareth the word of the kingdom, and understandeth it not, then cometh the wicked one, and catcheth away that which was sown in his heart. This is he which received seed by the way side. But he that received the seed into stony places, the same is he that heareth the word, and anon with joy receiveth it; Yet hath he not root in himself, but dureth for a while: for when tribulation or persecution ariseth because of the word, by and by he is offended. He also that received seed among the thorns is he that heareth the word; and the care of this world, and the deceitfulness of riches, choke the word, and he becometh unfruitful. But he that received seed into the good ground is he that heareth the word, and understandeth it; which also beareth fruit, and bringeth forth, some an hundredfold, some sixty, some thirty.

Because of the faith-testing environment the veil provides, Earth is the perfect stage for a world full of characters whose lives serve as active examples for all the host of heaven to observe and learn. Angels are watching us closely, and they are learning more about God than ever before—about why his ways are truly good and just (Ephesians 3:10; 1 Timothy 5:21; 1 Corinthians 4:9). They are watching us so intently, in fact, especially with respect to salvation, that they throw a big party whenever someone accepts Christ as his or her savior (1 Peter 1:10–12; Luke 15:7–10). They also observe with much dismay while we make all of our mistakes and then suffer the consequences of sin.

The host of heaven see the difference between those who rebel and those who repent and forgive. Billions of lives full of examples of selfishness versus selfless love are played out before them. From the perspective of the angels who have never sinned, the prospect of sinning might be a curious temptation, but thanks to us, as well as others in the universe who are suffering similar circumstances, they're learning that sin isn't all it's cracked up to be.

The people of Earth must use faith to see with the eyes of the heart, and they must make decisions in this dark world that will reveal the true intent of their hearts; but a day is coming when people won't have to be convinced that there really is a God. A day is coming when the veil will be taken away.

After Christ returns to Earth, Satan will be bound in chains in the underworld (Revelation 20:1–3). People born on Earth during that time will be able to see Jesus for themselves. In a sense, their faith won't be tested to the same degree as the people of Earth these days, which is why God gives an extra blessing to those who are tested in the darkness of these days. In light of this fact, we should stop complaining about how stinking hard life is and start appreciating the darkness we're immersed in because it gives us the opportunity to shine and have our confidence in God tested, unlike any other time in our entire existence in all of eternity ahead.

6

The Millennial Reign of Christ

In the Bible, heaven and Earth are always distinguished from each other. In the future, however, Earth will become more and more like the highest heaven until eventually it *becomes* the highest heaven, with the capital of the universe (New Jerusalem) relocated from its current planet to Earth, and set down in Israel, on the top of Mount Zion. But before the arrival of New Jerusalem, Jesus will reign on Earth in the natural Earth city of Jerusalem for one thousand years. This reign will commence immediately after Satan is apprehended and chained up in the bowels of Earth (Revelation 20:1–3).

Revelation 20:1–3

And I saw an angel come down from heaven, having the key of the bottomless pit and a great chain in his hand. And he laid hold on the dragon, that old serpent, which is the Devil, and Satan, and bound him a thousand years, and cast him into the bottomless pit, and shut him up, and set a seal upon him, that he should deceive the nations no more, till the thousand years should be fulfilled: and after that he must be loosed a little season.

During the millennial reign of Christ, many things on Earth will be the same, *but different*.

1. Earth Will Be the Same

As far as things being the same, the nature of Earth—the mountains, sky, plant and animal life—will all still be present. Earth will be the same real, physical place that it is now. People will continue to eat food as they do now because of the celebration of various feasts (feasts of Passover and unleavened bread—Ezekiel 45:21; first fruits—Ezekiel 44:30; Pentecost or Weeks—Ezekiel 46:9; trumpets—Ezekiel 44:5; day of atonement—Ezekiel 45–46; and tabernacles—Ezekiel 45:25, Zechariah 14:16–21). Because of the existence of food, agriculture will be one of the jobs people will have. Natural generations of people will have jobs just as they do now (agriculture and building are mentioned specifically—Isaiah 65:21–25). People will also tithe their incomes, since this has always been God's standard for a financial system. (Tithing was used before the law—Genesis 14:20, 28:22; under the law—Leviticus 27:30–33; Numbers 18:21; Nehemiah 10:37, 13:10–12; Proverbs 3:9–10; and after the fulfillment of the law through Jesus—Matthew 23:23; Romans 2:22; 1 Corinthians 9:7–18, 16:1–3; Hebrews 7.)

<u>Isaiah 65:21–25</u>

And they shall build houses, and inhabit them; and they shall plant vineyards, and eat the fruit of them. They shall not build, and another inhabit; they shall not plant, and another eat: for as the days of a tree are the days of my people, and mine elect shall long enjoy the work of their hands. They shall not labor in vain, nor bring forth for trouble; for they are the seed of the blessed of the LORD, and their offspring with them. And it shall come to pass, that before they call, I will answer; and while they are yet speaking, I will hear. The wolf and the lamb shall feed together, and the lion shall eat straw like the bullock: and dust shall be the serpent's meat. They shall not hurt nor destroy in all my holy mountain, says the LORD.

Government waste and corporate greed will no longer exist, and the many taxes people are forced to pay now will be replaced with a simple system of giving. In addition, many expenses will exist no longer (military, medicine, hospitals, cosmetics, crime prevention, and unhealthy addictive products such as drugs, tobacco, and alcohol). Earth will be filled with universal prosperity for all (Isaiah 65:21–24; Ezekiel 34:26; Micah 4:1–5; Zechariah 8:11–12, 9:16–17), but the same basic elements of society, including governmental structure and an economy, will still exist.

<u>Ezekiel 34:26</u>

And I will make them and the places round about my hill a blessing; and I will cause the shower to come down in his season; there shall be showers of blessing. And the tree of the field shall yield her fruit, and the earth shall yield her increase, and they shall be safe in their land, and shall know that I am the LORD, when I have broken the bands of their yoke, and delivered them out of the hand of those that served themselves of them.

Zechariah 8:12

For the seed shall be prosperous; the vine shall give her fruit, and the ground shall give her increase, and the heavens shall give their dew; and I will cause the remnant of this people to possess all these things.

I suspect artistic occupations will probably be much more prevalent in the kingdom of heaven, judging from the exquisite ornamentation included in every description of this realm. People will eventually find their natural gifts and skills and assume occupations that match their interests.

As previously stated, there will be cities with houses and streets (Revelation 21; John 14:1–3) and even public transportation, because the Bible speaks of highways that are traveled (Isaiah 11:16, 19:23–25, 35:7–8), and people making annual trips to the city of God (Isaiah 2:2–4, 52:7; Zechariah 8:20–23). Natural people probably won't have supernatural powers like those of the glorified saints, so it makes sense that they will use technology to supplement their needs. Considering all these facts, many things will be the same, and nearly every hobby and activity there is on Earth now will probably be available in those days as well (subtracting sinful elements, of course).

Isaiah 11:16 (bold emphasis added)

And there shall be an highway for the remnant of his people, which shall be left, from Assyria; like as it was to Israel in the day that he came up out of the land of Egypt.

Isaiah 19:23–25 (bold emphasis added)

In that day shall there be **a highway out of Egypt to Assyria**, and the Assyrian shall come into Egypt, and the Egyptian into Assyria, and the Egyptians shall serve with the Assyrians. In that day shall Israel be the third with Egypt and with Assyria, even a blessing in the midst of the land: Whom the LORD of hosts shall bless, saying, Blessed be Egypt my people, and Assyria the work of my hands, and Israel mine inheritance.

Zechariah 8:20–23 (bold emphasis added)

Thus says the LORD of hosts; It shall yet come to pass, that **there shall come people, and the inhabitants of many cities: And the inhabitants of one city shall go to another, saying, Let us go speedily to pray before the LORD**, and to seek the LORD of hosts: I will go also. Yea, **many people and strong nations shall come to seek the LORD of hosts in Jerusalem**, and to pray before the LORD. Thus says the LORD of hosts; In those days it shall come to pass, that ten men shall take hold out of all languages of the nations, even shall take hold of the skirt of him that is a Jew, saying, We will go with you: for we have heard that God is with you.

There will be natural generations of people on Earth for all eternity; people will have families and children forever. Initially they will begin as the survivors of the great tribulation. Following the battle of Armageddon, one of the first assignments of the glorified saints will be to go out into the world and notify all the people of the world of the new kingdom of Christ on Earth (Isaiah 2:2–4, 52:7; Zechariah 8:23). The inhabitants of the kingdom of Christ on Earth, referenced in Matthew 5:5 and Revelation 21:1–22:6, will contain glorified saints (1 Thessalonians 4:17; 1 Corinthians 15:52), but natural generations of people are specifically mentioned as well (Isaiah 11:6–7, 65:20, 65:25).

Isaiah 2:3 (bold emphasis added)

And many people shall go and say, Come ye, and let us go up to the mountain of the LORD, to the house of the God of Jacob; and he will teach us of his ways, and we will walk in his paths: for **out of Zion shall go forth the law, and the word of the LORD from Jerusalem.**

1 Thessalonians 4:17—these are glorified saints

Then we which are alive and remain shall be caught up together with them in the clouds, to meet the Lord in the air: and so shall we ever be with the Lord.

1 Corinthians 15:52—these are glorified saints

In a moment, in the twinkling of an eye, at the last trump: for the trumpet shall sound, and the dead shall be raised incorruptible, and we shall be changed.

Isaiah 11:6–9—these children are from natural generations of people (bold emphasis added)

The wolf also shall dwell with the lamb, and the leopard shall lie down with the kid; and the calf and the young lion and the fatling together; and **a little child shall lead them.** And the cow and the bear shall feed; their young ones shall lie down together: and the lion shall eat straw like the ox. **And the sucking child shall play on the hole of the asp, and the weaned child shall put his hand on the cockatrice' den.** They shall not hurt nor destroy in all my holy mountain: for the earth shall be full of the knowledge of the LORD, as the waters cover the sea.

The Millennial Reign of Christ 219

It won't be until after the millennial reign of Christ that death won't exist anymore (Isaiah 25:8; 1 Corinthians 15:26; Revelation 20:14).

> Isaiah 25:8
>
> He will swallow up death in victory; and the Lord GOD will wipe away tears from off all faces; and the rebuke of his people shall he take away from off all the earth: for the LORD hath spoken it.
>
> 1 Corinthians 15:26–28
>
> The last enemy that shall be destroyed is death. For he hath put all things under his feet. But when he says all things are put under him, it is manifest that he is excluded, which did put all things under him. And when all things shall be subdued unto him, then shall the Son also himself be subject unto him that put all things under him, so that God may be all in all.

The people of Earth will continue to advance in love and prosperity forever, become more technologically advanced with each passing generation, and eventually spread out into the cosmos in search of new territory. They will be in the kingdom of Christ, which shall increase forever, so naturally they will eventually need to spread to other planets (Isaiah 9:6–7; Daniel 7:13–14; Micah 4:1–3; Zechariah 9:10, 14:1–21; Revelation 11:15).

> Isaiah 9:6–7* (bold emphasis added)
>
> For unto us a child is born, unto us a son is given: and the government shall be upon his shoulder: and his name shall be called Wonderful, Counselor, The mighty God, The everlasting Father, The Prince of Peace. **Of the increase of his government and peace there shall be no end**, upon the throne of David, and upon his kingdom, to order it, and to establish it with judgment and with justice from henceforth even for ever. The zeal of the LORD of hosts will perform this.
>
> *An Earth full of immortal people who continue to reproduce will eventually become populated to the point that expansion beyond Earth will become necessary. This is especially the case concerning a kingdom that increases forever, as the scripture above states.*
>
> Daniel 7:13–14 (bold emphasis added)
>
> I saw in the night visions, and, behold, one like the Son of man came with the clouds of heaven, and came to the Ancient of days, and they brought him near before him. And there was given him dominion, and glory, and a kingdom, that all people, nations, and languages, should serve him: **His dominion is an**

everlasting dominion, which shall not pass away, and his kingdom that which shall not be destroyed.

In essence, the way many non-Christian futurists perceive the future of the world is exactly the way it will be, provided one has the most optimistic outlook and confidence in the achievements of humanity one could possibly have. The humanistic psychology view of the future, as advocated in *Star Trek*, is very close to what the future on Earth will be like, with one huge exception—Gene Roddenberry didn't mention Jesus, the one who will make it all possible!

2. Earth Will Be Different

Everything will be the same, but at the same time, everything will be very different. After the millennial reign, there will be mountains on Earth, but Mount Zion will be unlike any mountain that anyone has ever seen. It will be 1,500 miles high (Revelation 21:16), and with the atmosphere as clean and crisp as it will be, this mountain will probably be visible from hundreds of miles away. Even more intriguing is the fact the atmosphere of Earth will have to be altered to accommodate this mountain, because if such a mountain were on Earth nowadays, satellites would be colliding with it starting at the 200 mile mark, less than one-seventh of the way up![1]

Also concerning changes in the natural realm of Earth, there will still be days and nights forever (though nighttime will not be perceived within the city of New Jerusalem when it comes), but everything will be seven times brighter than it currently is during the day (Isaiah 30:26; 60:18–22). As stated earlier, the sun being larger in the first heaven might have something to do with this.

Isaiah 30:26

Moreover **the light of the moon shall be as the light of the sun, and the light of the sun shall be sevenfold, as the light of seven days**, in the day that the LORD binds up the breach of his people, and heals the stroke of their wound.

Because of the intensity of the increased light, normal human vision will probably be altered as well, so people will be able to function in such a bright place, rather than groping around blindly. One might think so much light would simply be perceived as a blurry brightness, but this is far from true. The increased light will add more color and vividness to Earth. Consider the vivid colors in the movie *Finding Nemo*; the reason this movie is so colorful in some scenes is that over two hundred virtual lights were used in the lighting effects (this was described on the extra DVD that comes with the movie).[2]

As for the animals on Earth in those days, carnivores will no longer exist. For lions to eat straw, and other predators such as wolves to not attack sheep, they have to be dramatically altered (Isaiah 11:6–7, 65:25).

Moreover one of the biggest changes on Earth beginning with the millennial reign of Christ will be the fact that Earth will once again be a glorified realm, open to the heavens and freely accessible to and from other glorified realms. *Star Trek*, which I've alluded to so frequently, gives a faint glimmer of this reality. Glorified beings of all sizes, shapes, and degrees of power and authority will be traveling to and from Earth. Some angels will most likely even live on Earth in New Jerusalem when it comes, because they're with the Lord in the third heaven right now, and many of them serve God directly at his throne (all visions of angels and beasts in Revelation chapters 4 through 22 were in the third heaven). Because these beings currently live in New Jerusalem, and this city will be transported to Earth, there's no reason to believe that they will be evicted from their homes when this happens.

The first heaven in the atmosphere above Earth will be accessible to the glorified saints in the days of the millennial reign of Christ, and they will be going back and forth between that realm and Earth. They will also be visiting the highest heaven as well (glorified saints have their homes in the highest heaven—John 14:2; yet they will be ruling over Earth during the millennium, which will require frequent travel—Revelation 1:4–6, 5:10; then after the New Jerusalem is moved to Earth, they will live on Earth—Revelation 21:1–2:6).

John 14:2–3* (bold emphasis added)

In my Father's house are many mansions: if it were not so, I would have told you. I go to prepare a place for you. And if I go and prepare a place for you, I will come again, and receive you unto myself; that where I am, there ye may be also.

*The Father's house is currently in the highest heaven, so the glorified saints will live in the highest heaven, but will be required to govern the Earth as kings and priests as described in the scriptures below.

Revelation 1:4–6* (bold emphasis added)

John to the seven churches which are in Asia: Grace be unto you, and peace, from him which is, and which was, and which is to come; and from the seven Spirits which are before his throne; and from Jesus Christ, who is the faithful witness, and the first begotten of the dead, and the prince of the kings of the earth. Unto him that loved us, and washed us from our sins in his own blood, and **hath made us kings and priests unto God and his Father**; to Him be glory and dominion for ever and ever. Amen.

To be kings and priests, glorified saints will have to spend time in the Earth, even though their homes will be in the highest heaven during the millennial reign of Christ.

Revelation 5:10

And hast made us unto our God kings and priests: and we shall reign on the earth.

Revelation 21:2–3*

And I John saw the holy city, New Jerusalem, coming down from God out of heaven, prepared as a bride adorned for her husband. And I heard a great voice out of heaven saying, Behold, the tabernacle of God is with men, and he will dwell with them, and they shall be his people, and God himself shall be with them, and be their God.

Once New Jerusalem is transplanted to the Earth at the conclusion of the millennial reign of Christ, the glorified saints won't have to travel to Earth from the highest heaven anymore because Earth will be the new highest heaven, where God's throne will be.

3. The First Assignment Given to the Saints

As crazy as it sounds, many people on Earth following the establishment of the new kingdom of Christ will not even realize that the Antichrist had come and gone, or that Jesus will be the new king of Earth. Multitudes of people in remote regions all over the world will probably know that something monumental has happened, because the wrath of God poured out on the world in the end times will probably touch everyone to some extent, but they won't know it had anything to do with the Antichrist, or Jesus's return to Earth.

The scriptures teach that many people will not see the glory of Christ or his person the day he comes to Earth (Isaiah 66:19).

Isaiah 66:19

And I will set a sign among them, and I will send those that escape of them unto the nations, to Tarshish, Pul, and Lud, that draw the bow, to Tubal, and

Javan, to the isles afar off, that have not heard my fame, neither have seen my glory; and they shall declare my glory among the Gentiles.

Every eye will see him only within the vicinity of Jerusalem, when Jesus returns at the second advent of Christ (Matthew 24:27–31; Revelation 1:7). After Christ begins his reign, missionaries and rulers from Jerusalem (the resurrected saints will take over the governments of the world) will go from Jerusalem to tell of Christ reigning in Zion. When the nations learn of this, they will say, "Let us go up to the mountain of the Lord." (Isaiah 2:2–4; 52:7; Zechariah 8:23). That will be an incredible sight to see—countless numbers wandering across Earth, going to see the Lord in person. Imagine the excitement in these people.

4. Sin and Death Still Present During the Millennial Reign of Christ

During the millennial reign of Christ, sin and death will still exist on Earth. There will be two classes of people on Earth in those days—those glorified into an angelic nature, and those who will still be natural generations of humans, just as Adam and Eve were. The natural generations of humans will be those who did not take the mark of the beast, and were not killed by either the wrath of God or the Antichrist.

While the glorified saints will be immortal as angels, as discussed in an earlier chapter, the natural generations of humans will not be; they will only be pseudoimmortal as Adam and Eve were, depending on the tree of life for their continued existence. Scripture states that the days of man will be lengthened (Isaiah 65:20; Zechariah 8:4), and those who stay faithful to God won't die, just as Adam and Eve would've never died had they never sinned. These natural generations of people will eventually be translated into angels if they remain faithful to God, but there will always be natural generations of people on Earth. Natural generations of people will reproduce forever, because each generation will go through a reproductive stage prior to being translated into angelic form.

Natural generations of people will be permitted to eat from the tree of life and receive healing by eating the leaves of the tree of life (Revelation 22:2). They won't be immortal as angels are, however, until they are translated. If this was not the case, there would be no way to explain natural generations of people on Earth (Isaiah 2:1–2, 11:6–7, 65:25; Zechariah 14:16–21; Revelation 11:15, 20:1–10; Matthew 25:31–46), while at the same time, resurrected saints will be like the angels which are in heaven, which do not marry, and therefore do not have children (Matthew 22:30; Mark 12:25; Luke 20:35; Revelation 21:16).

It will still be possible for sin to exist in the kingdom of Christ during the millennial reign because, when Satan is loosed after one thousand years (at the end of the millennial reign of Christ), he will manage to deceive many people into following him one last time. Since sin will be possible (on Earth, with the exception of New Jerusalem), if people sin on Earth in those days, God will most likely deal with them the same way he did in the past with Adam and Eve. Those who are stupid enough to sin and refuse to repent will not be allowed to eat from the trees of life anymore, just as Adam and Eve were kept from the tree of life in the Garden of Eden. This is why Isaiah 65:20 states that the aging process will be different for sinners during that time.

Isaiah 65:20

> There shall be no more thence an infant of days, nor an old man that hath not filled his days: for the child shall die an hundred years old; but the sinner being an hundred years old shall be accursed.

Unbelievably, there will still be people who will hunger for a sinful lifestyle, even after living in paradise for a thousand years. Near the end of the thousand-year reign of Christ, they will make a very foolish decision.

5. The Eternal Future of Life on Earth

At the conclusion of the thousand-year reign of Christ, God will allow Satan to be loosed one last time. The reasoning behind this has to do with purging the last remnant of sinners from Earth who managed to survive the days of the great tribulation. Many who will not take the mark of the beast will survive, but some among these survivors will not be devoted to God. They probably won't even like living in his kingdom because their attraction to sin and darkness will make them discontent with even the most blissful of places.

Generation after generation of sinful people will accumulate on Earth, and by the time the millennial reign of Christ nears its end, Earth will be full of utterly selfish, immoral, uncaring, unloving, prideful, and arrogant people. This has to be the case, or Satan wouldn't be able to deceive so many of them. Scripture states clearly that there will be an immense army raised up by Satan, when he is loosed for the last time. They will converge on New Jerusalem, thinking that they might actually accomplish something, and then suddenly an ocean of fire will be dumped on their heads, raining down from the sky (Revelation 20:9). In a sense, it will be anticlimactic. Boom! All dead, game over.

Revelation 20:9

And they went up on the breadth of the earth, and compassed the camp of the saints about, and the beloved city: and fire came down from God out of heaven, and devoured them.

Shortly after this event, God will dust off his hands, then commence with the completion of a heavens-and-earth remodel job that's been pending for a long time. The first heaven and the first Earth will pass away (Revelation 21:1—the wording in the Hebrew for the phrase "pass away" refers to a reordering, rearranging, or changing, rather than a destruction). I believe what this entails is the reintegration of the multidimensional structure of the universe, or multiverse, into its original unified structure. I derive this conclusion from the scriptural description of events, as well as scientific and logical reasons, which I will further divulge.

Concerning the scriptural description of events, the changes described in 2 Peter 3:10 indicate a fundamental altering of the atomic structure of matter itself, the magnitude of which could easily refer to a dimensional alteration. The terminology used specifically suggests a *melting* of the heavens and Earth, which fits with the idea of an integration of Earth into the first heaven. Adding to this is the clear statement made in Revelation 21:1: the *first heaven* and the *first Earth* will be passing away.

Revelation 21:1 (bold emphasis added)

And I saw a new heaven and a new earth: **for the first heaven and the first earth were passed away**; and there was no more sea.

2 Peter 3:10 (bold emphasis added)

But the day of the Lord will come as a thief in the night; in the which the heavens shall pass away with a great noise, and **the elements shall melt with fervent heat**, the earth also and the works that are therein shall be burned up.

At the same time this melting occurs, scripture speaks of changes within Earth, on Earth, and in the sky, which is where the first heaven is located. All these changes occur simultaneously.

In the sky, New Jerusalem (the capital of the universe) will most likely be planted in the first heaven first because everything that comes to Earth from higher levels of glory (the highest heaven, namely) always passes through the first heaven first—just as the angels did in the book of Daniel (Daniel 8:15–27, 10:5–21). Then, after New Jerusalem is planted in the first heaven, the Earth below will be raised up and integrated into the first heaven. From the perspective of the inhabitants of Earth, it will look as though New Jerusalem is descending down onto Earth (Revelation 21:2–3), when in fact the opposite will be true: Earth will be ascending, both physically, and in glory.

After this ascension takes place, Earth will be considered the new third heaven because the capital of the universe will be relocated from the planet where it's currently at to Earth.

In addition to this city made of unearthly materials (transparent gold) appearing to float down from the sky, the changes on the surface of Earth will be momentous. There will be seven times more light (Isaiah 30:26, 60:18–22), and a great deal more land will be available due to the fact that there won't be any more oceans (Revelation 21:1–2, yet there will still be an abundance of water—Isaiah 30:23–25, 33:20–21, 35:6–7, 41:17–18, 49:10; Ezekiel 34:26, 47:1–12; Zechariah 14:8; Joel 3:18). Also, land will be restored to unparalleled fruitfulness (Isaiah 35:1–10, 55:12, 55:13; Ezekiel 36:8–12; Joel 2:18–27, 3:17–21; Amos 9:13–15), and the mountains will be much larger. The mountains of Earth will be so large, in fact, that Earth will have to be increased in size in order to accommodate them.

As for alterations made in the core of Earth, hell will be cast into the lake of fire (Revelation 20:14), and the lake of fire most likely won't be located within Earth anymore; the heart of Earth will therefore be purified. The Bible actually doesn't give any clear statements as to the location of the lake of fire. It wouldn't make any sense, however, to say that hell will be removed from where it already exists (in the center of Earth), and then cast into another place (the lake of fire), which would actually be the same place, if both hell and the lake of fire exist within the center of Earth. It is true that the lake of fire will be visible from God's throne room (Isaiah 66:22–24; Revelation 14:9–11), but that doesn't mean that it's located in the center of Earth. Probably everything will be visible from God's throne room, if God so chooses it.

Concerning the scientific support for the idea that Earth will merge into the first heaven, take note of the colossal Mount Zion previously mentioned, located in the center of New Jerusalem. This mountain will be 1,500 miles high! That's *miles*, not feet. The existence of this immense mountain suggests that Earth will have to be increased in size. Only a larger planet with a higher altitude atmosphere could accommodate such a mountain. If a mountain this size were on

Earth today, satellites would be colliding with it, and there wouldn't be any air or gravity before getting close to the middle of it. I'm not even sure if it's possible to have such a mountain so disproportionately large compared to the planet that houses it without defying the laws of physics. It's true that God can do anything, but he created the universe with the laws of physics in mind, and the only time he bends or breaks those laws is to prove a point. He does not simply break them just because he can. His creation generally remains consistently confined to the laws of physics.

Such a mountain could probably exist on a much larger planet, as I mentioned earlier. (Remember the onion theory. The first heaven is the outer layer of the onion, which is much bigger than the Earth of today, which is a middle layer.) It only makes sense that God would increase the size of Earth to accommodate Mount Zion, as well as the other extremely large mountains that will be on Earth in those days. Earth will be a home for many more people and even a bunch of angels at that time, so making it a bigger planet will serve several functions.

Finally, now that I've disclosed both the scriptural description of events and the scientific reasons behind why I think the multidimensional universe will be transformed into a singular universe again, the last reason, being a logical conclusion, remains. Logically speaking, the entire purpose for having a multidimensional universe will be moot after Judgment Day; there will be no more need to divide realms in order to prevent cursed beings from entering glorified realms because all cursed beings will be cast into the lake of fire and all cursed realms will be restored to a state of glory. There will be no corruption within Earth, on Earth, above Earth, or in the vastness of the cosmos. The heart of Earth will no longer serve as a prison for dissidents, but it will be restored; as for the first heaven, there will be no need to keep it separate from the glorified Earth.

In summation, Earth will be the new third heaven, containing God's throne, where God will rule the universe for all eternity. Earth will exist in openness and harmony with all the other worlds in the vast universe, into exceeding levels of glory that surpass every imagination.

Part V
The Grand Conclusion

I started this book with a list of questions, all of which highlight the primary theme of this book—that extraterrestrial life is a reality spoken of in the Bible, and it will play a key role in the end times. By now, the answer to every question addressed should be understood.

Aliens—who or what are they?

Aliens are exactly what most people have been saying they are: intelligent lifeforms not native to Earth. They dwell in the outer atmosphere of Earth (the first heaven), and among the stars; these regions are the multilayered dimensions of the universe (multiverse), which the Bible calls the heavens. In biblical terminology, some of them are angels (both faithful and fallen), and some have not become angels yet, or perhaps may never be angels. They can all be referred to as the host of heaven. The Bible also refers to intelligent life in the heavens with terms such as "stars" (a term referring to both the heavenly beings and their dwellings—Job 38:4–7; Revelation 12:4), "creatures" (Ezekiel 1:5–7; Revelation 5:13–14), and powers, principalities, and rulers in heavenly places (Ephesians 6:12).

What do aliens want?

This depends on whether they are angels or not, and whether they have been deceived or not. If they are angels faithful to God, they serve God, help humanity, learn about salvation, and spread the knowledge of salvation to all who may benefit from it. If they are fallen angels, they serve Satan either directly or indirectly by serving their own selfish motives. The primary aspect of all their actions concerns deception, for this is their greatest power. They can cause far more destruction in their war against God and his elect through deception than through any other means. The only way Satan can truly destroy a soul is through deception rather than physical destruction.

In relation to this deception, there is a very well-defined satanic agenda, of which I may have outlined a few key aspects in this book. Concerning alien abductions, it's likely that genetic experimentation is related to the development of a new breed of Nephilim, of which the Antichrist will be the capstone achievement. Genetic experimentation may serve numerous other purposes as well, such as building up their antibodies to our diseases, so their visitations will be less hazardous for them. They could even be building biological weapons. I find the prospect of good angels using genetics to help us to be less likely, in any scenario, simply because God's faithful angels are noted in scripture as having superior healing powers derived from a simple touch.

Concerning alien life in general, I suspect the reality of alien life will be related to a new religion espoused by the Antichrist. The heart of all the deceptions derived from this new belief system will be to achieve one objective: to point people away from the identity of Christ as the savior, and the only begotten Son of God the Father, Creator of all things.

As for beings who are not angels, yet who have not sinned, they are most likely assisting angels. Beings in this category are as humans would have been, had Adam and Eve never sinned. An easy way to identify them is that they are pseudoimmortal, and they reproduce, so they're good and glorified to a certain extent, but they aren't angels. They are also more likely to need technology to assist them because they aren't as powerful as those who are glorified into angelic form.

Finally, those who are not angels, but have sinned as the humans of Earth have, will fall on both sides of the fence concerning what they want. Some will seek out God and hunger for the knowledge of salvation, while others could care less about God, in which case they will assist Satan, either directly or indirectly through serving their own selfish motives. The things that beings in this category could want vary greatly. Some might be scientists using people like cattle, as some people speculate. Then again they might penetrate society and use their advantages to reap the easily obtainable pleasures of this world.

Why do all these beings come to Earth?

As I already stated, for many of them, the reason behind their visits has to do with salvation. Those aligned with God are seeking salvation or learning more about and spreading the knowledge of salvation and all the aspects of redemption, and those not aligned with God are attempting to block the efforts of spreading the knowledge of salvation, either directly or indirectly. Cursed beings exploiting humanity to obtain the pleasures of this world, for example, would indirectly block the efforts of spreading the knowledge of salvation, because their motives would be sinful, and it would be easy for Satan to manipulate them to do his bidding.

The genetic experimentation being conducted by some of them may very well be related to the construction of a new breed of Nephilim, which—again—will be another attempt at blocking the efforts of salvation through a new and unique deception for the modern scientific age.

Have aliens been on Earth in the past?

For many distraught beings, Earth was their original home, and it was destroyed, which is one of the sources of their extreme consternation. They were here before Adam and Eve, they were here before and after the days of Noah (Genesis 6:4), and they will be here in open contact in the future, as in the days of Noah (Matthew 24:37; Luke 17:26).

Are aliens among us right now?

Entire civilizations of faithful and fallen angels live right above our heads in the sky, dimensionally shifted from our perception, in a realm the Bible refers to as the first heaven.

Also, knowing that a biblical equivalent of the term "alien" can be "angel," scripture clearly teaches that these beings are visibly among us even now—often right before our very eyes, as undercover agents in disguise (Hebrews 13:2).

As for alien abductions (the unpleasant ones), these are another example of aliens among us. Alien abductions are rare because they're generally conducted by beings who are taking a big risk. Earth is protected by the prayers of the saints (James 5:16; and specifically the great restrainer of the Antichrist—2 Thessalonians 2:6–8). The prayers of saints actually serve as orders given to angelic forces operating in the heavens, and wars are fought whenever boundary restrictions are violated and fallen angels penetrate into Earth. That's why abductees have been known to report that their abductors immediately return them when abductees start praying to Jesus for help. (I've personally had a number of abductees tell me about this.)

Battles in the first heaven go the other way too; they are also fought by fallen angels in an attempt to prevent the good angels from getting through to us (Daniel 10:13). This is why prayers appear to be hindered at times—because they are!

What role will extraterrestrial life play in the future of humanity?

In the future, following the rapture of the church, the force field of prayer (restraining power) currently protecting Earth will be removed (2 Thessalonians 2:6–8), and open contact with alien life will most likely immediately follow. First those in the heavens who are aligned with Satan will come (Daniel 8:10–12; Matthew 24:15, 24:29–31; Luke 21:24–28; and Revelation 12:3–5), and then God's faithful angels will penetrate the world in open contact as well—preaching the Gospel (Revelation 14:6–10).

In addition to angels preaching the Gospel, the two Old Testament prophets, Enoch and Elijah, will also return from heaven and preach the Gospel. A massive conversion of 144,000 Jewish Christians (Messianic Jews), and many new Gentile (non-Jewish) Christians will be supernaturally empowered by God more than any generation before them (Joel 2:29–32; Acts 2:17–18; John 14:12).

As always, people will have true revelation given side by side with deception, and the condition of their hearts will motivate them to choose one way or the other. An unprecedented revival will encompass the globe. At the same time, deception and sin will be greater on Earth than ever in the past. In essence, Earth will be prepared for a harvest, to separate the wheat from the tares; in other words, the majority of the people on Earth will be clearly polarized in their alignment with either God or Satan.

During the reign of the Antichrist, otherworldly battles will ensue, and the Antichrist will prevail against these otherworldly beings (Daniel 8:10–12, Matthew 24:15, 24:29–31; Luke 21:24–28; Revelation 12:3–5). The otherworldly combatants in these battles will most likely be warring factions within Satan's kingdom. I say this because Satan will not prevail against the armies of God when they finally arrive with Jesus at the second advent of Christ. It won't matter to the Antichrist that his kingdom is divided, however, because he'll use his victory in these battles to elevate his position on Earth, and eliminate the opposition within his own kingdom.

One reason for the victories of the Antichrist in the future—even over extraterrestrial attacks—will be his alliance with other extraterrestrials who will help him (Daniel 11:39). With such an alliance, the technology made available to him will be extremely advanced, even to the degree that the creation of true artificial intelligence will become a reality (Revelation 13:14–16).

The otherworldly allies of the Antichrist, headed by a fallen angel by the name of the Destroyer (Revelation 9:11, Hebrew "Abaddon," Greek "Apollyon"), will perpetuate the deceptions of the Antichrist by giving credence to his grandiose claims. The Antichrist will claim to be of heavenly origin, and a destined leader of the world; his otherworldly allies will then step forward and say it's true (Revelation 13:11–14).

On the one hand, the Antichrist's claim of heavenly descent will be partially true (making his deception particularly powerful). The Antichrist might be a Nephilim (fathered by a fallen angel), linking him with the heavenly realm. The part that the Antichrist (and those aligned with his plan) will leave out, however, is the fact that the Nephilim are cursed of God, and their very existence reeks of the highest order of abominations (angels who sired them were imprisoned—1

Peter 3:19; 2 Peter 2:4; Jude 6–7; and the entire world was destroyed because of the Nephilim—Genesis 6). Therefore the respect, admiration, and worship he will claim to deserve as a divine right will be the biggest, fattest lie ever told.

What can the people of Earth do during this tumultuous time in the future?

The whole reason I have written this book is primarily for the people left behind after the rapture of the church—those who will face a chaotic world of warring nations, otherworldly beings, and supernatural displays beyond imagination.

What will you do when the prophecies spoken of in this book begin to take place? Perhaps nearly a billion people have recently disappeared off the planet, and you are looking for answers in this book called *Aliens and the Antichrist* because reality has become something you weren't expecting.

Start by seeking out the truth, and turn to the Appendix in the back of this book. It is definitely possible to be saved after the rapture of the church!

Those who taste the freedom Jesus spoke of will never accept anything less (John 8:32, 10:10). For those who are truly saved and know it in their hearts, the sky may roll up like a scroll, oceans and seas may spill across the continents, and the mountains can melt like wax, but nothing will take away that peace from God that transcends all understanding (Philippians 4:7).

> John 8:32
>
> And ye shall know the truth, and the truth shall make you free.
>
> John 10:10
>
> The thief cometh not, but for to steal, and to kill, and to destroy: I am come that they might have life, and that they might have it more abundantly.
>
> Philippians 4:7
>
> And the peace of God, which passes all understanding, shall keep your hearts and minds through Christ Jesus.

Accept Christ—he is the only way! In his own words, he is the way, the truth, and the life—there is no way to the Father except by him (John 14:6).

> John 14:6
>
> Jesus said unto him, I am the way, the truth, and the life: no man cometh unto the Father, but by me.

I also have a few practical suggestions for the rough, yet tremendously exciting road ahead (for those reading this after the rapture of the church).

- Identify deceptions when they arrive. This is achieved by being aware of the warnings in the Bible, many of which are documented in this book, and especially cultivating a right relationship with God. Concerning specific warnings in the Bible, take note of a few cut-and-dried scriptures warning about end time deceptions, deceivers, and false prophets.

Deuteronomy 13:1–4 (bold emphasis added)

If there arise among you a prophet, or a dreamer of dreams, and gives thee a sign or a wonder, And the sign or the wonder come to pass, whereof he spoke unto thee, saying, Let us go after other gods, which thou hast not known, and let us serve them; thou shalt not hearken unto the words of that prophet, or that dreamer of dreams: for the LORD your God proves you, to know whether ye love the LORD your God with all your heart and with all your soul. Ye shall walk after the LORD your God, and fear him, and keep his commandments, and obey his voice, and ye shall serve him, and cleave unto him.

Deuteronomy 18:20–22 (bold emphasis added)

But **the prophet, who shall presume to speak a word in my name, which I have not commanded him to speak, or that shall speak in the name of other gods, even that prophet shall die**. And if thou say in your heart, how shall we know the word which the LORD hath not spoken? **When a prophet speaks in the name of the LORD, if the thing follows not, nor come to pass, that is the thing which the LORD hath not spoken, but the prophet hath spoken it presumptuously: thou shalt not be afraid of him**.

1 John 4:1–4 (bold emphasis added)

Beloved, believe not every spirit, but **try the spirits whether they are of God**: because many false prophets are gone out into the world. Hereby know ye the Spirit of God: **Every spirit that confesses that Jesus Christ is come in the flesh is of God: And every spirit that confesseth not that Jesus Christ is come in the flesh is not of God: and this is that spirit of antichrist**, whereof ye have heard that it should come; and even now already is it in the world.

According to these scriptures, there are essentially two tests for prophets. First, if they don't have any divine ability to back up what they're saying—they predict things that *don't* happen as they say—then they are clearly not from God. Second, even if they do exercise a certain degree of accuracy in foretelling future events, but their doctrine they preach is false, then again, they are clearly false prophets. Two examples of false doctrine are given in scripture. First, that Jesus Christ is not the messiah he said he was, and second, any word spoken in the name of other gods. The second test will be the one in which the Antichrist will publicly break with his religion advocating a strange god. For more examples of various deceptions, conduct research on other religions, such as Islam (which claims that Jesus was not the Son of God and that he did not rise from the dead). Also, simply pick up practically any book from the New Age section of any local bookstore and browse the gambit of philosophical belief systems that describe the number of ways people can go about perfecting themselves, essentially becoming gods. Deception is everywhere, but truth is everywhere too; pray to God out of a hope that he is love, and he will guide into all truth.

Psychics frequently fail the first test, then attribute their misses to fluctuations in Earth's magnetic field, or make other excuses. Prophets, on the other hand, need no excuses for being wrong because they should never be wrong about what they are prophesying. The distinction between a true prophet and one who claims to have the gift of prophecy but actually doesn't was a fatal distinction in Daniel's time because King Nebuchadnezzar didn't accept any excuses for being wrong. If they were wrong about their prophetic abilities, then they were lying to him—the king—and were thus ordered to be executed.

- Knowing whose side to be on is just the beginning. The next step is to never accept the mark of the beast, or the name or the number of the beast, when this becomes a requirement in the nations the Antichrist will rule (Revelation 13:17). Not everyone in the world in those days will be forced to make this decision, but for those who are, if they don't want to go to hell, they had better not take the mark, the name, or the number of the beast. Only those who will not take this mark will be resurrected and escape the second death (Revelation 20:14–15).

- Have courage and be strong in the faith—especially when it becomes a life-and-death situation (Revelation 12:11; 14:12). Don't ever give up hope, even if it looks as if torture may come; God either rescues his saints or empowers them to endure the unimaginable. Many books have been written which document the amazing feats of thousands of martyrs throughout the millennia. The Jesus Freaks book I mentioned earlier, along with its second volume companion, is full of miraculous stories of bravery, sacrifice, and God's direct intervention. In any case, God will never leave us or forsake us (Deuteronomy 31:6). Even if it comes to loved ones being tortured, one should never give in. Always remember that evil lies in the hands of the one committing the acts, not in the hands of one being manipulated through such evil coercion.

- The days following the rapture of the church will be days in which Christians will display remarkable supernatural power unlike ever before (Joel 2:29–32; Acts 2:17–18; John 14:12). Christians should be ready, willing, and eager to accept this power because they're going to need it, and they should never limit what God can do through those who believe in him.

As I've already stated, the primary beneficiaries of the material presented in this book are those who remain on Earth after the rapture of the church, but this book serves a purpose to people before the rapture of the church as well. I have written it not as an intellectual quandary, but to give people a taste of the reality of the kingdom of heaven. It's a real, physical, tangible place!

The spiritual dimension of reality doesn't just need to be understood; it needs to be known in the heart so that it affects the way we think, the way we live, and the direction of all of our relationships. As Christians, our lives should be a witness of the glory within us—a perfection we pursue with our commitment to God. The fact that we know what's going on around us—that we're in a war that constantly puts us to the test, and that our every action is being evaluated—should place a deeper meaning on our lives. The sky above our heads and the stars that sparkle in the vast cosmos at night are the realms we can look at and see with our own eyes, and they are realms we will one day explore.

The religious people of Jesus's day (scribes and Pharisees) would have intellectual debates with Jesus about the scriptures, which they knew a great deal about. But then, in the same breath, they would accuse him of violating the Sabbath when he healed someone (Matthew 12:10; Mark 3:2–6; Luke 6:6–11; 13:10–14, and many others). Didn't they realize what just happened? If that weren't enough, they plotted to kill Christ when they couldn't win their little intellectual squabbles. Apparently all their intellectual knowledge amounted to squat because their hearts were so far away from God, they couldn't see God's kingdom displayed right before their eyes in glaring physical manifestations. Those scribes and Pharisees back then aren't any different from many people of today.

This same differentiation can be made between the mind and the heart concerning this book. The format and style of this book is generally intellectual in nature, with the application of scriptures, logic, and deductive reasoning, but the goal is not to pass on the accumulation of knowledge. The goal of this book is to wake people up and inflame their imaginations with images of the kingdom of Heaven and the days that lie ahead; images of the vast worlds that lay beyond, of the immensity of the war in which we are entrenched, and the eternal spiritual reality that surrounds us all. The embers of our complacent hearts must be kindled, and we should all live in a perpetual state of anticipation, never being satisfied with the distractions of this fading world.

Just as the apostle Paul said, without love (that portion of our hearts connected to God), all the knowledge and mysteries of all creation amount to nothing (1 Corinthians 13). Only love remains in this world as the truest representation of God that will survive forever. So let's take this truth and live by it, dream our dreams of the frontier awaiting us, and walk with a vision of the future that will carry us through this small, transitory tunnel of mortality.

*** **THE END** ***

APPENDIX

The Gospel

WHY DID GOD CREATE?

The heart of the Creator of all things is love (1 John 4:16). Because God is love, he decided to create a beautiful universe and fill it with life. Why? So he can be with and love his children (1 Corinthians 2:9; Ephesians 1:4–5; John 1:1–2, 1:14). Love grows, and love enjoys loving relationships. God didn't want this relationship to disappear someday. He wants to spend eternity with his loving children, so he created his children as eternal beings (Acts 24:15; Daniel 12:2; Genesis 1:26–27).

WHY DOES EVIL EXIST?

Of course, in order for his children to love him, they must freely *choose* to love him. Therefore, God decided to create life with free will (Genesis 2:16–17; Deuteronomy 30:19–20; Joshua 24:15).

Because of the existence of free will, the opportunity to choose not to love and obey God exists. This can be a sad fact because many will choose not to love and obey God, but it must be so, in order for love to bear true meaning. This is why evil exists. God allows evil to exist for a short season (on the scale of eternity), in order for all to choose him with their own free will. Once this choice has been made throughout all creation, evil will pass away (Revelation 20:14–15).

THE MEANING OF LIFE

That's the meaning of life, by the way—to choose to love and obey God, and have an eternal loving relationship with him. Only those who have true love in their hearts can understand this. Once this choice is made, the meaning of life grows to incorporate revealing this truth to others, so that they may also choose to love and obey God and have an eternal loving relationship with him (1 Corinthians 2:9; Ephesians 1:4–6; John 1:1–2, 1:14).

WHAT HAPPENS WHEN FREE WILL IS USED FOR EVIL?

Unfortunately, because of the existence of free will, sin has occurred on Earth, beginning in Genesis 3:6. What's worse, sin has a spreading effect once it is turned loose. There is no human means to stop its corruption. Sin was turned loose on Earth with the very first man and woman (Adam and Eve) and has spread throughout the entire Earth since that time (Romans 3:23, 5:12–15).

Because God is perfect (Psalms 18:30), there is no tolerance for sin; the penalty of sin is death (Genesis 2:17; Romans 5:12). God defines death as returning to the dust of the Earth (Genesis 3:19) and calls it a curse, known as the last enemy (1 Corinthians 15:26). In essence any rejection of God's rule is sin. It began with Adam and Eve and has since been endorsed by all humanity (Romans 5:12). Therefore, it appears there is a great dilemma, because after sin was introduced into the world, all have been born into sin, living in bodies with a sinful nature. All are penalized with death, even from birth.

GOD'S METHOD OF DEALING WITH SIN

God knew that humanity needed a means of overcoming sin. That's why the first thing he said about sin concerned his means of correcting it (Genesis 3:14–15). His Son, Jesus, is the answer, who is first mentioned as the seed of the woman in Genesis 3:14–15. God decided that since humanity doesn't have the power to deal with sin and death (despite all the religions and doctrines in the world that claim otherwise), he had to take care of this problem himself. He therefore embedded himself inside the womb of a human female by the name of Mary and became a human being. The Creator of all things became human (John 1:1–2, 1:14; Colossians 2:9–10). His name was Jesus and still is. He lived a life of purity and obedience to God—a life of love, healing, and forgiveness. Because of his willpower, he was able to live perfectly, without sin (Hebrews 4:15). He then ended his life on Earth by sacrificing himself for all (Romans 5:8–10; 1 Peter 3:18). Again by his power, he defeated death by rising from the dead (John 10:17–18, Romans 8:10–11).

GOD'S CALLING

God offers the free gift of salvation to all who will ask. He calls upon all of us to turn from our sinful ways and trust in what he has done for us. We can do nothing to remove our guilt before him. Doing good things doesn't remove our sin, and since all are sinners, nothing we can do can undo that; it's only by the mercy of God that we can be saved through what he has done (Ephesians 2:8–10). The gift of salvation is eternally rewarded, but those who refuse to repent and turn to God will have no place in his kingdom. Therefore whoever spurns God's offer will suffer his wrath in the judgment to come, of which the Bible clearly warns. This terrifying prospect (2 Thessalonians 1:7–10) is a reality that Jesus spoke much of, warning people of their fate.

GOD'S JUSTICE

God is perfect justice (Psalms 19:7), perfect mercy (2 Corinthians 1:3–5), and all-powerful (Romans 1:19–20). Jesus's sacrifice is the best example of all of these attributes. The fact that he died and paid the penalty of sin is justice. God accepts sacrifice, and allowed his Son, Jesus Christ, to be sacrificed to atone for all of the sin in the world. The price had to be paid, but he picked up the tab! Whatever sins anyone has ever committed—past, present, or future—have all been atoned for. All people have to do is ask, and believe.

GOD'S MERCY

The fact that Jesus died, yet Jesus is without sin is mercy. He didn't have to die; he is blameless, yet he died for all who would ask him to take their place (Romans 5:8; 1 Peter 3:18). This was the requirement in order for his sacrifice to atone for the sin of the world. If he were a sinner himself, he'd only be getting the punishment he deserved, but because he is perfect, his crucifixion allowed him to be made worthy to forgive all who would come to him and ask his forgiveness.

GOD'S POWER

Jesus's perfect life (Hebrews 4:15), his miracles, and especially his resurrection, all demonstrate the fact that he is all-powerful. His only limitation was to confine himself to the will of his Father. He died to atone for sin, and rose from the dead to prove that the atonement was not without meaning. He has demonstrated his power over death, and he will give that power to all who ask him for forgiveness of sins and salvation.

WHY JESUS IS THE ONLY WAY TO ETERNAL LIFE

When God the Father allowed his Son to be sacrificed, it wasn't easy. Would it be easy for anyone to allow his or her child to be murdered? He doesn't take his Son's death lightly, despite the fact that Jesus was resurrected. God stands outside of time and space, and can see his Son's death as if it were present—for all eternity! He can hear the cries of his Son, "Father, why have you forsaken me" (Matthew 27:46; Mark 15:34), for all eternity. This is his means of saving humanity, and there is *no* other way. If there were, he would've taken it, because neither he nor his Son desired the brutal murder of the cross. This is serious business, and it's why God allows for no other means of salvation. We should be glad there is no other way, however, because his way is the surest and the best—though it is sad that Jesus had to be crucified in order to make it happen. Nobody has to go through any rituals or punishments. All people have to do is lay aside their pride and stop trying to reach God by their own strength. All they have to do is look to Jesus, because he is the way, the truth, and the life; there is no way to the Father, except by him (John 14:6).

HOW MIGHT ONE HAVE ETERNAL LIFE?

Salvation is a free gift. There's only one thing that people have to do to receive it. Ask Jesus for forgiveness of their sins. He forgives all! The only unforgivable sin is the blasphemy of the Holy Spirit, which is achieved essentially by refusing God's offer and not asking for forgiveness. Of course one must have enough faith to ask such a question. This is a problem for many because they simply don't perceive what faith actually is.

Let me make this easy. Anyone reading this who fits in this category needs to simply ask him, Jesus—to forgive you of your sins and save you—and hope he's real, if this is all you can do. I assure you, in his infinite mercy, you are saved if you do this. Put your hope in him, and he will not let you down. You don't even have to change your life before you come to him. He'll take care of those details later on. Just stop right now—and send up a prayer to him.

There are other ways of rewording the act of asking God for salvation. For example, Acts 20:20–21 states that one must have "Repentance toward God and faith toward our Lord Jesus Christ." Repentance means a complete change of heart and mind regarding sin—in which a person agrees with God about his or her sin and wants to live a life pleasing to him. Faith in Jesus Christ entails accepting who he is, "The Son of the living God," that "Christ died for the ungodly," and that he conquered death through his resurrection (1 Corinthians 15:1–4; 15:21–22). All of these Christian phrases simply boil down to believing in God enough to ask for forgiveness of sins from his Son, Jesus Christ, and allowing him to work in people to mold them into his likeness. He really carries the brunt of the workload—mainly because people are so spiritually weak and puny.

It's that easy. God bless you!

Scriptures Referenced in Appendix

1 John 4:16

And we have known and believed the love that God hath to us. God is love; and he that dwells in love dwells in God and God in him.

1 Corinthians 2:9

But as it is written, Eye hath not seen, nor ear heard, neither have entered into the heart of man, the things which God hath prepared for them that love him.

Ephesians 1:4–5

According as he hath chosen us in him before the foundation of the world, that we should be holy and without blame before him in love: Having predestinated us unto the adoption of children by Jesus Christ to himself, according to the good pleasure of his will.

John 1:1–2, 14*

In the beginning was the Word, and the Word was with God, and the Word was God. The same was in the beginning with God…And the Word was made flesh, and dwelt among us (and we beheld his glory, the glory as of the only begotten of the Father), full of grace and truth.

God became flesh to be intimately linked to us and to save us.

Acts 24:15*

And have hope toward God, which they themselves also allow, that there shall be a resurrection of the dead, both of the just and unjust.

All spirits are eternal, whether they are obedient and faithful to God or not. Everlasting damnation isn't God's desire, for he is not willing that any should perish, but it exists because it is a byproduct of free will. The fact that all spirits are eternal, however, is evident in that they exist for eternity, whether they are saved with God, or eternally separated from God.

Daniel 12:2*

And many of them that sleep in the dust of the earth shall awake, some to everlasting life, and some to shame and everlasting contempt.

All spirits are eternal, whether they are obedient and faithful to God or not. Everlasting damnation isn't God's desire, for he is not willing that any should perish, but it exists because it is a byproduct of free will. The fact that all spirits are eternal, however, is evident in that they exist for eternity, whether they are saved with God or eternally separated from God.

Genesis 1:26–27

And God said, Let us make man in our image, after our likeness: and let them have dominion over the fish of the sea, and over the fowl of the air, and over the cattle, and over all the earth, and over every creeping thing that creeps upon the earth. So God created man in his own image, in the image of God created he him; male and female created he them.

Genesis 2:16–17*

And the LORD God commanded the man, saying, Of every tree of the garden thou might freely eat: But of the tree of the knowledge of good and evil, thou shalt not eat of it: for in the day that thou eat thereof thou shalt surely die.

While it isn't explicitly stated that humans have free will, it is implied by the fact that God created the tree of the knowledge of good and evil, and gave Adam and Eve the choice of whether to obey him and not eat it, or to disobey him, and eat it.

Deuteronomy 30:19–20*

I call heaven and earth to record this day against you, that I have set before you life and death, blessing and cursing: therefore choose life, that both thou and thy seed may live: That thou might love the LORD thy God, and that thou might obey his voice, and that thou might cleave unto him: for he is thy life, and the length of thy days: that thou might dwell in the land which the LORD swore unto thy fathers, to Abraham, to Isaac, and to Jacob, to give them.

While it isn't explicitly stated that humans have free will, it is implied by the fact that people obviously have the ability to choose life—and to love and obey God, as it is mentioned in this passage of scripture.

Joshua 24:15*

And if it seem evil unto you to serve the LORD, choose you this day whom ye will serve; whether the gods which your fathers served that were on the other side of the flood, or the gods of the Amorites, in whose land ye dwell: but as for me and my house, we will serve the LORD.

While it isn't explicitly stated that humans have free will, it is implied by the fact that people obviously have the ability to choose life—and to love and obey God, as it is mentioned in this passage of scripture.

Revelation 20:14–15

And death and hell were cast into the lake of fire. This is the second death. And whosoever was not found written in the book of life was cast into the lake of fire.

1 Corinthians 2:9

But as it is written, Eye hath not seen, nor ear heard, neither have entered into the heart of man, the things which God hath prepared for them that love him.

Ephesians 1:4–6

According as he hath chosen us in him before the foundation of the world, that we should be holy and without blame before him in love: Having predestinated us unto the adoption of children by Jesus Christ to himself, according to the good pleasure of his will, To the praise of the glory of his grace, wherein he hath made us accepted in the beloved.

John 1:1–2, 14

In the beginning was the Word, and the Word was with God, and the Word was God. The same was in the beginning with God…And the Word was made flesh, and dwelt among us (and we beheld his glory, the glory as of the only begotten of the Father), full of grace and truth.

Genesis 3:6

And when the woman saw that the tree was good for food and that it was pleasant to the eyes, and a tree to be desired to make one wise, she took of the fruit thereof, and did eat, and gave also unto her husband with her; and he did eat.

Romans 3:23

For all have sinned, and come short of the glory of God.

Romans 5:12

Wherefore, as by one man sin entered into the world, and death by sin; and so death passed upon all men, for that all have sinned...

Psalms 18:30

As for God, his way is perfect: the word of the LORD is tried: he is a buckler to all those that trust in him. For who is God save the LORD?

Genesis 2:17

But of the tree of the knowledge of good and evil, thou shalt not eat of it: for in the day that thou eat thereof thou shalt surely die.

Romans 5:12–15

Wherefore, as by one man sin entered into the world, and death by sin; and so death passed upon all men, for that all have sinned: For until the law sin was in the world: but sin is not imputed when there is no law. Nevertheless death reigned from Adam to Moses, even over them that had not sinned after the similitude of Adam's transgression, who is the figure of him that was to come. But not as the offence, so also is the free gift. For if through the offence of one many be dead, much more the grace of God, and the gift by grace, which is by one man, Jesus Christ, hath abounded unto many.

Genesis 3:19

In the sweat of thy face you will eat bread, until thou return unto the ground; for out of it wast thou taken: for dust thou art, and unto dust you will return.

1 Corinthians 15:26

The last enemy that shall be destroyed is death.

Romans 5:12–15

Wherefore, as by one man sin entered into the world, and death by sin; and so death passed upon all men, for that all have sinned: For until the law sin was in the world: but sin is not imputed when there is no law. Nevertheless death reigned from Adam to Moses, even over them that had not sinned after the similitude of Adam's transgression, who is the figure of him that was to come. But not as the offence, so also is the free gift. For if through the offence of one many be dead, much more the grace of God, and the gift by grace, which is by one man, Jesus Christ, hath abounded unto many.

Genesis 3:14–15

And the LORD God said unto the serpent, Because thou hast done this, thou art cursed above all cattle, and above every beast of the field; upon thy belly you will go, and dust you will eat all the days of thy life: And I will put enmity between thee and the woman, and between thy seed and her seed; it shall bruise thy head, and thou shalt bruise his heel.

John 1:1–2, 14

In the beginning was the Word, and the Word was with God, and the Word was God. The same was in the beginning with God…And the Word was made flesh, and dwelt among us (and we beheld his glory, the glory as of the only begotten of the Father), full of grace and truth.

Colossians 2:9–10

For in him dwells all the fullness of the Godhead bodily. And ye are complete in him, which is the head of all principality and power.

Hebrews 4:15

For we have not an high priest which cannot be touched with the feeling of our infirmities; but was in all points tempted like as we are, yet without sin.

Romans 5:8–10

But God commended his love toward us, in that, while we were yet sinners, Christ died for us. Much more then, being now justified by his blood, we shall be saved from wrath through him. For if, when we were enemies, we were reconciled to God by the death of his Son, much more, being reconciled, we shall be saved by his life.

1 Peter 3:18

For Christ also hath once suffered for sins, the just for the unjust, that he might bring us to God, being put to death in the flesh, but quickened by the Spirit.

John 10:17–18

Therefore doth my Father love me, because I lay down my life, that I might take it again. No man takes it from me, but I lay it down of myself. I have power to lay it down, and I have power to take it again. This commandment have I received of my Father.

Romans 8:10–11

And if Christ be in you, the body is dead because of sin; but the Spirit is life because of righteousness. But if the Spirit of him that raised up Jesus from the dead dwell in you, he that rose up Christ from the dead shall also quicken your mortal bodies by his Spirit that dwells in you.

Ephesians 2:8–10

For by grace are ye saved through faith; and that not of yourselves: it is the gift of God: Not of works, lest any man should boast. For we are his workmanship, created in Christ Jesus unto good works, which God hath before ordained that we should walk in them.

2 Thessalonians 1:7–10

And to you who are troubled rest with us, when the Lord Jesus shall be revealed from heaven with his mighty angels, In flaming fire taking vengeance on them that know not God, and that obey not the gospel of our Lord Jesus Christ: Who shall be punished with everlasting destruction from the presence of the Lord, and from the glory of his power; When he shall come to be glorified in his saints, and to be admired in all them that believe (because our testimony among you was believed) in that day.

Psalms 19:7

The law of the LORD is perfect, converting the soul: the testimony of the LORD is sure, making wise the simple.

2 Corinthians 1:3–5

Blessed be God, even the Father of our Lord Jesus Christ, the Father of mercies, and the God of all comfort; Who comforted us in all our tribulation, that we may be able to comfort them which are in any trouble, by the comfort wherewith we ourselves are comforted of God.

Romans 1:19–20

Because that which may be known of God is manifest in them; for God hath showed it unto them. For the invisible things of him from the creation of the world are clearly seen, being understood by the things that are made, even his eternal power and Godhead; so that they are without excuse.

Romans 5:8

But God commended his love toward us, in that, while we were yet sinners, Christ died for us.

1 Peter 3:18

For Christ also hath once suffered for sins, the just for the unjust, that he might bring us to God, being put to death in the flesh, but quickened by the Spirit.

Hebrews 4:15

For we have not an high priest which cannot be touched with the feeling of our infirmities; but was in all points tempted like as we are, yet without sin.

Matthew 27:46

And about the ninth hour Jesus cried with a loud voice, saying, Eli, Eli, lama sabachthani? That is to say, My God, my God, why hast thou forsaken me?

Mark 15:34

And at the ninth hour Jesus cried with a loud voice, saying, Eloi, Eloi, lama sabachthani? Which is, being interpreted, My God, my God, why hast thou forsaken me?

John 14:6

Jesus said unto him, I am the way, the truth, and the life: no man cometh unto the Father, but by me.

Acts 20:20–21

And how I kept back nothing that was profitable unto you, but have showed you, and have taught you publicly, and from house to house, Testifying both to the Jews, and also to the Greeks, repentance toward God, and faith toward our Lord Jesus Christ.

1 Corinthians 15:1–4

Moreover, brethren, I declare unto you the gospel which I preached unto you, which also ye have received, and wherein ye stand; By which also ye are saved, if ye keep in memory what I preached unto you, unless ye have believed in vain. For I delivered unto you first of all that which I also received, how that Christ died for our sins according to the scriptures; And that he was buried, and that he rose again the third day according to the scriptures.

1 Corinthians 15:21–22

For since by man came death, by man came also the resurrection of the dead. For as in Adam all die, even so in Christ shall all be made alive.

Notes

Foreword

1. Brad Steiger, *The UFO Abductors* (New York, New York: The Berkley Publishing Group, 1988), 202.

2. *Wikipedia Encyclopedia Online*, s.v. "Panspermia," http://en.wikipedia.org/wiki/Panspermia (accessed December 4, 2005).

Overview

1. Fenis Dake, *God's Plan for Man* (Lawrenceville, Georgia: Dake Publishing Inc., 1977), 80. Sources throughout these notes pointing to this book might include page numbers that change in successive reprints, because this book was first printed in 1949, and has been reprinted twenty-two times since then. If the page number isn't relevant to the text for this entry, refer to the index, and look up the word "Heavens."

2. Rick Meyers, "Equipping Ministries Foundation, e-Sword Bible software," April 7, 2000, http://www.e-sword.net/bibles.html; *Enhanced Strong's Lexicon* (Oak Harbor, WA: Logos Research Systems Inc., 1995).

Part 1

1. Komal Birguni, "How many stars are there in the sky?" *Ask Yahoo!* http://ask.yahoo.com/20010810.html (accessed December 27, 2005); T. Joseph W. Lazio, "How many galaxies in the Universe?" *Internet FAQ Archives: Online Education*, http://www.faqs.org/faqs/astronomy/faq/part8/section-4.html (accessed December 27, 2005); CNN, Sydney, Australia, "Star survey reaches 70 sextillion," *CNN*, http://www.cnn.com/2003/TECH/space/07/22/stars.survey/(accessed December 27, 2005).

Chapter 2

1. Rick Meyers, "Equipping Ministries Foundation, e-Sword Bible software," April 7, 2000, http://www.e-sword.net/bibles.html; *Enhanced Strong's Lexicon* (Oak Harbor, WA: Logos Research Systems Inc., 1995).

2. Ibid.

3. Fenis Dake, "Biblical Proof That the Beast Out of the Abyss is 'The Prince of Grecia,'" *God's Plan for Man* (Lawrenceville, Georgia: Dake Publishing Inc., 1977), 991–922. If the page number isn't relevant to the text for this entry, refer to the index, and look up the phrase "Prince of Grecia."

4. W. Raymond Drake, "Gods and Spacemen in Greece and Rome," (Sphere, London 1976), 115–116, quoted in Ufoevidence.org, "329 BC: Alexander the Great records two great 'flying shields'" *UFO Evidence*, http://www.ufoevidence.org/cases/case491.htm (accessed January 22, 2006).

5. Ufoevidence.org, "329 BC: Alexander the Great records two great 'flying shields'" *UFO Evidence*, http://www.ufoevidence.org/cases/case491.htm (accessed January 22, 2006).

6. Michael S. Heiser, *The Facade* (Superior, Colorado: Superior Books, 2001), 289–290.

7. Dr. David P. Stern, "The Discovery of Atoms and Nuclei," *NASA funded Web site*, http://www-spof.gsfc.nasa.gov/stargaze/Ls7adisc.htm (accessed March 19, 2006); Antony C. Wilbraham and others, *Chemistry* (Upper Saddle River, New Jersey: Prentice Hall, Inc., 2005), 102.

8. Amanda Gefter, "Is string theory in trouble?" *NewScientist.com*, http://www.newscientist.com/channel/fundamentals/quantum-world/mg18825305.800 (accessed December 27, 2005).

9. David Deutsch, Homepage, http://www.qubit.org/people/david/index.php?path=Home (accessed December 29, 2005).

10. Noah Shachtman, "Real-Life Hyperspace Drive," *Military.com*, http://www.defensetech.org/archives/002065.html (accessed January 12, 2006).

11. Ian Johnston, "Welcome to Mars Express: Only a Three Hour Trip," *Scotsman.com, Scottish News, Sports & Scottish Headlines Direct from Scotland*, http://news.scotsman.com/scitech.cfm?id=16902006 (accessed January 12, 2006).

12. Ibid.

13. Neal Singer, "Another dramatic climb toward fusion conditions for Sandia Z accelerator," *Sandia National Laboratories*, http://www.sandia.gov/media/z290.htm (accessed January 12, 2006).

14. Trent Brandon, "Poltergeists," *Zerotime Paranormal Website*, http://www.zerotime.com/ghosts/polter.htm (accessed December 29, 2005). This referenced Web site gives many examples of poltergeists moving objects.

15. Rosemary Ellen Guiley, "Haunting," in *Harpers Encyclopedia of Mystical & Paranormal Experience* (San Francisco: HarperSanFrancisco, 1991), 253–254.

16. John Edward, Homepage, http://www.johnedward.net (accessed December 29, 2005).

17. Charles T. Tart, "Psychic's Fears of Psychic Powers," *The Journal of the American Society for Psychical Research* 80, no. 3 (July 1986): 279–92.

18. Rosemary Ellen Guiley, "Haunting," in *Harpers Encyclopedia of Mystical & Paranormal Experience* (San Francisco: HarperSanFrancisco, 1991), 253–254.

19. Dennis William Hauk, "Gettysburg National Military Park," *Hauntedhouses.com*, http://hauntedhouses.com/states/pa/house5.htm (accessed December 31, 2005); Phil Keller, "Gettysburg, PA." *The Shadowlands*, http://theshadowlands.net/famous/gettysburg.htm (accessed December 31, 2005).

20. Beth Scott and Michael Norman, *Haunted Heartland: Lincoln and the Supernatural* (New York, New York: Dorset Press, 1985), 68–74.

21. Arthur Myers, *The Ghostly Register* (Chicago, Illinois: Contemporary Books, 1986), 242–246. Ghostly wagons in this story were never seen, but they were heard, along with numerous cattle, horses, cowboys cracking whips and

yelling, and a baby crying. A number of people visiting this region reported these strange noises that were very loud, yet no one could see any visible trace of what was causing them.

22. Ibid, 300–304.

23. Rosemary Ellen Guiley, "Thought-forms," in *Harpers Encyclopedia of Mystical & Paranormal Experience* (San Francisco: HarperSanFrancisco, 1991), 616–618.

24. *To Catch a Killer*, DVD, directed by Eric Till (Chatsworth, California: Image Entertainment, September 21, 1999).

25. Rosemary Ellen Guiley, "Shroud of Turin," in *Harpers Encyclopedia of Mystical & Paranormal Experience* (San Francisco: HarperSanFrancisco, 1991), 550–552.

26. Ibid, "Out-Of-Body Experiences," 419–423.

27. *Readers Digest: The Worlds Last Mysteries* (Pleasantville, New York, Montreal: The Reader's Digest Association Inc., 1979), 177.

28. Wade Cox, "Mysticism, chapter 1 Spreading the Babylonian Mysteries," *Christian Churches of God*, http://www.ccg.org/english/s/b7_1.html (accessed December 28, 2005).

29. *Readers Digest. The Worlds Last Mysteries* (Pleasantville, New York, Montreal: The Reader's Digest Association Inc., 1979), 168–179.

30. *Christian Answers Network Bible Encyclopedia*, s.v. "Nimrod: Who was he? Was he godly or evil?" (by Dr. David P. Livingston) http://christiananswers.net/q-abr/abr-a002.html (accessed January 21, 2006).

31. Pritchard, J., *Ancient Near Eastern Texts and the Old Testament*, 3rd edition (Princeton: University Press, 1969), quoted in *Christian Answers Network Bible Encyclopedia*, s.v. "Nimrod: Who was he? Was he godly or evil?" (by Dr. David P. Livingston) http://christiananswers.net/q-abr/abr-a002.html (accessed January 21, 2006).

32. *Christian Answers Network Bible Encyclopedia*, s.v. "Nimrod: Who was he? Was he godly or evil?" (by Dr. David P. Livingston) http://christiananswers.net/q-abr/abr-a002.html (accessed January 21, 2006).

33. Frank E. Carlisle, "Solving the Riddle of Ezekiel's Wheels: The Chariots of the Cherubim/The Chariots of God," *UFOInfo.com*, http://ufoinfo.com/news/ezekielswheels.shtml (accessed December 31, 2005).

34. Terry Melanson, "UFOs & the Cult of ET: The Phantasmagorical Manipulation," *Illuminati Conspiracy Archive*, http://www.conspiracyarchive.com/UFOs/UFOs_Aliens_Contactees.htm (accessed January 2, 2006); Brad Steiger. *The Fellowship: Spiritual Contact Between Humans and Outer Space Beings* (New York, New York: Ivy Books, 1988), 170.

35. Carol R. Ember and Melvin Ember, "Chapter 18, Applied Anthropology," *Cultural Anthropology, Sixth Edition* (Englewood Cliffs, New Jersey: Prentice-Hall Inc., 1990), 337–354.

36. Sophy Burnham, *A Book of Angels* (New York, New York: Ballentine Books, 1990), 35–37.

Part 2

1. Edward L. Wright, "Age of the Universe," *USLA Division of Astronomy and Astrophysics*, http://www.astro.ucla.edu/~wright/age.html (accessed December 30, 2005).

2. *Wikipedia Encyclopedia Online*, s.v. "Gender-neutral Language," http://en.wikipedia.org/wiki/Gender-neutral_language (accessed December 28, 2005).

Chapter 3

1. William K. Purves and others, *Life: The Science of Biology, Fourth Edition* (Sunderlan, Massachusetts: Sinauer Associates Inc., 1995), 646.

2. Brenda Flynn, "Index," *Watcher's Website*, http://www.mt.net/~watcher/new.html (accessed January 2, 2006).

3. Brenda Flynn, "Asteroids, Comets, Rahab & Mars, Cherubim & the Megaliths of Cydonia, The Stones of Fire and Pre-Adamite Civilizations,"

Watcher's Website, http://www.mt.net/~watcher/stones.html (accessed January 3, 2006).

4. Pat Dasch and Allan Treiman, "Ancient Life on Mars???" *Lunar and Planetary Institute*, http://www.lpi.usra.edu/publications/slidesets/marslife/index.shtml (accessed January 2, 2006).

5. Malin Space Science Systems, "The Face on Mars," *Malin Space Science Systems*, http://barsoom.msss.com/education/facepage/face.html (accessed January 3, 2006).

6. Brenda Flynn, "Asteroids, Comets, Rahab & Mars, Cherubim & the Megaliths of Cydonia, The Stones of Fire and Pre-Adamite Civilizations," *Watcher's Website*, http://www.mt.net/~watcher/stones.html (accessed January 3, 2006).

7. Ibid.

8. Ibid.

9. Ibid.

10. Tariq Malik, "Most Americans Believe Alien Life is Possible, Study Shows," *Space.com*, http://www.space.com/news/050531_alienlife_survey.html (accessed January 3, 2006).

11. John B. Alexander, "The Alexander UFO Religious Crisis Survey: The Impact of UFOs and their Occupants on Religion," *National Institute for Discovery Science*, http://www.nidsci.org/articles/alexander/response_analysis.html (accessed January 2, 2006).

12. Rick Meyers, "Equipping Ministries Foundation, e-Sword Bible software," April 7, 2000, http://www.e-sword.net/bibles.html; *Enhanced Strong's Lexicon* (Oak Harbor, WA: Logos Research Systems Inc., 1995).

13. Michael S. Heiser, *The Facade* (Superior, Colorado: Superior Books, 2001), 261–270.

14. Ibid, 272.

15. Fenis Dake, *God's Plan for Man* (Lawrenceville, Georgia: Dake Publishing Inc., 1977), 480–481. Look up "Trinity" in the index to obtain all the page numbers for this source; there are twenty-five subtopics scattered throughout the book that relate to the Trinity.

16. Brenda Flynn, "UFOS AND THE BIBLE, Aliens: The Angelic Conspiracy," *Watcher's Website*, http://www.mt.net/~watcher/ufos.html (accessed January 4, 2006); Gil Bar and Barry Chamish, Israeli UFO Research, http://members.tripod.com/~ufoisrael/index.html (accessed December 12, 2005).

17. Barry Chamish, "UFO Wave in Israel," *Great Dreams Website*, http://www.greatdreams.com/chamish.htm (accessed December 16, 2005).

18. Ibid, "Chapter 5, Israel Awakens to the Invasion."

19. Rick Meyers, "Equipping Ministries Foundation, e-Sword Bible software," April 7, 2000, http://www.e-sword.net/bibles.html; *Enhanced Strong's Lexicon* (Oak Harbor, WA: Logos Research Systems Inc., 1995).

20. Ibid.

21. Chris Stassen, "The Age of the Earth," *The Talk Origins Archive: Exploring the Creation/Evolution Controversy*, http://www.talkorigins.org/faqs/faq-age-of-earth.html (accessed January 6, 2006).

22. William K. Purves and others, *Life: The Science of Biology, Fourth Edition* Sunderlan, Massachusetts: Sinauer Associates Inc., 1995), 646.

23. Ibid. While much of the information in this book points to scientific data that indicates the Earth is billions of years old, it may not actually be that old. It might only be thousands of years old, as creation scientists contend, but this point is irrelevant in light of the theories I argue in this book. I do believe that the Earth is millions of years older than Adam and Eve, for several reasons I point out in this book, but the Earth doesn't have to be billions of years old as most mainstream secular scientists believe.

24. Dr. Walt Brown, "Online Edition, In the Beginning: Compelling Evidence for Creation and the Flood (7th Edition)," *Center for Scientific Creation*, http://www.creationscience.com/onlinebook/FAQ312.html (accessed January 5, 2006).

25. Rick Meyers, "Equipping Ministries Foundation, e-Sword Bible software," April 7, 2000, http://www.e-sword.net/bibles.html.

26. Ibid.

27. William K. Purves and others, *Life: The Science of Biology, Fourth Edition* (Sunderlan, Massachusetts: Sinauer Associates Inc., 1995), 636–640.

28. Ibid.

29. Ibid.

30. Ibid.

31. Dennis R. Petersen, *Unlocking the Mysteries of Creation: The Explorer's Guide to the Awesome Works of God* (El Dorado, California: Creation Resource Publications, 2002), 30. This book, which supports Creationism, refutes carbon dating, and argues that believing in the old Earth theory supports evolution, but my argument is that it doesn't matter how old the Earth is in order for the Bible to be accurate; Marvin L. Lubenow, *Bones of Contention: A Creationist Assessment of Human Fossils* (Grand Rapids, Michigan: Baker Books, 1992), 86–119; Marvin L. Lubenow, People Fossils: The Evidence for Human Evolution, http://www.thebigmystery.com/human_evolution.htm (accessed January 10, 2006).

32. Marvin L. Lubenow, People Fossils: The Evidence for Human Evolution, http://www.thebigmystery.com/human_evolution.htm (accessed January 10, 2006). The human arm fragment listed here is but one example among many others that have been found throughout the years, discussed later in this book.

33. Russell Grigg, "Pre-Adamite man: were there human beings on Earth before Adam?" *Answers in Genesis: Upholding the Authority of the Bible from the Very First Verse*, http://www.answersingenesis.org/creation/v24/i4/humans.asp (accessed January 10, 2006).

34. William K. Purves and others, *Life: The Science of Biology, Fourth Edition* (Sunderlan, Massachusetts: Sinauer Associates Inc., 1995), 636–640.

35. David Getz, *Frozen Man* (New York, New York: Scholastic Inc., 1994), 19–20.

36. Stephen Wagner, "The 10 Most Puzzling Ancient Artifacts: The Grooved Spheres," *About.com/Paranormal Phenomena*, http://paranormal.about.com/library/weekly/aa011402a.htm (accessed January 7, 2006).

37. Vera Solovieva, "The Map of The Creator," *PRAVDA.Ru Russian News and Analysis*, http://english.pravda.ru/main/2002/04/30/28149.html (accessed December 24, 2005).

38. Stephen Wagner, "The 10 Most Puzzling Ancient Artifacts: 33 Million Year Old Mortar and Pestle Mystery," *About.com/Paranormal Phenomena*, http://paranormal.about.com/od/ancientanomalies/ (accessed January 7, 2006).

39. Steve Byerly, OOPARTS, http://byerly.org/whatifo.htm (accessed January 11, 2006).

40. Dennis R. Petersen, "Do Artifacts of Very Ancient and Even pre-Flood Technology Exist?" *Unlocking the Mysteries of Creation: The Explorer's Guide to the Awesome Works of God*, 204–205, 224–225.

41. *Readers Digest: The Worlds Last Mysteries* (Pleasantville, New York, Montreal: The Reader's Digest Association Inc., 1979), 14.

42. Ibid.

43. Ibid, 11–27; Ellie Crystal, Ellie Crystal's Metaphysical and Science Website, http://www.crystalinks.com/atlantis4.html (accessed January 5, 2006). While I don't agree with a great deal of Ellie Crystal's Web site, I validated much of the research she put into the section covering Atlantis by cross-referencing it with my highly reputable Reader's Digest book. The most notable archeological findings pointing to Atlantis, in my opinion, are those found submerged off the island of North Bimini in the Bahamas. Information about these ancient ruins proliferate libraries, and the Internet.

44. *Readers Digest: The Worlds Last Mysteries* (Pleasantville, New York, Montreal: The Reader's Digest Association Inc., 1979), 14.

45. *Wikipedia Encyclopedia Online*, s.v. "Furlong," http://en.wikipedia.org/wiki/Furlong (accessed January 4, 2006).

46. WiseGEEK, "What are the World's Tallest Mountains?" *wiseGEEK: Clear Answers for Common Questions*, http://www.wisegeek.com/what-are-the-worlds-tallest-mountains.htm (accessed January 12, 2006).

47. Lisa Gardiner, Windows to the Universe: Big, Bigger, Biggest, http://www.windows.ucar.edu/tour/link=/our_solar_system/relative_size.html (accessed January 12, 2006).

48. Brad Harrub, Ph.D., "Computer-Generated Neanderthals," *Apologetic's Press on the World Wide Web*, http://www.apologeticspress.org/articles/2062 (accessed January 12, 2006).

49. William K. Purves and others, *Life: The Science of Biology, Fourth Edition* (Sunderlan, Massachusetts: Sinauer Associates Inc., 1995), 636–640.

50. Brad Harrub, Ph.D, "Computer-Generated Neanderthals," *Apologetic's Press on the World Wide Web*, http://www.apologeticspress.org/articles/2062 (accessed January 12, 2006).

Chapter 4

1. Voice of the Martyrs, "Home Page," *Voice of the Martyrs presents Persecution.com*, http://www.persecution.com (accessed January 15, 2006).

2. DC Talk, *Jesus Freaks: DC Talk and The Voice of the Martyrs—Stories of Those Who Stood For Jesus, the Ultimate Jesus Freaks* (Bloomington, Minnesota: Bethany House Publishers, January 1, 1999), inside flap of front cover.

3. Voice of the Martyrs, "Restricted Nations Profiles, China," *Voice of the Martyrs presents Persecution.com*, http://www.persecution.com (accessed January 15, 2006).

4. Voice of the Martyrs, "Restricted Nations Profiles, Pakistan," *Voice of the Martyrs presents Persecution.com*, http://www.persecution.com (accessed January 15, 2006).

5. Franco Cavazzi, Illustrated History of the Roman Empire, http://www.roman-empire.net (accessed January 15, 2006).

6. Chuck Missler, "Our Blessed Hope," *Koinonia House Online: Bringing the world into focus through the lens of Scripture*, http://www.khouse.org/articles/

2002/444/ (accessed January 15, 2006); Chuck Missler, *The Rapture: Christianity's Most Preposterous Belief*, Audio Cassette (Coeur d' Alene, Idaho: Koinonia House, December 10, 2002).

7. Fenis Dake, *God's Plan for Man* (Lawrenceville, Georgia: Dake Publishing Inc., 1977), 882–887. If the page numbers aren't relevant to the text, refer to the index, and look up the term "man child."

8. Brian Schwertley, "Is the Pretribulation Rapture Biblical?" *Grace Online Library*, http://www.graceonlinelibrary.org/articles/full.asp?id=9%7C21%7C40 (accessed January 12, 2006).

9. Antonio Casolari, "The Systematic Idiocy: The War," *Vency's Site: Liberty, Knowledge, Reason*, http://www.vency.com/wars.html (accessed January 12, 2001).

10. Jack Fellman, "Eliezer Ben-Yehuda and the Revival of Hebrew (1858–1922)," *Jewish Virtual Library*, http://www.jewishvirtuallibrary.org/jsource/biography/ben_yehuda.html (accessed January 17, 2006).

11. Central Intelligence Agency (CIA), "Israel," *CIA: The World Fact Book*, http://www.cia.gov/cia/publications/factbook/geos/is.html (accessed January 15, 2006).

12. Ami Isseroff, "A Brief History of Israel and Palestine and the Conflict," *MidEast Web Gateway*, http://www.mideastweb.org/briefhistory.htm (accessed January 17, 2006).

13. Ibid.

14. Lambert Dolphin, "Moving Towards a Third Jewish Temple," *The Temple Mount in Jerusalem*, http://www.templemount.org/tempprep.html (accessed January 17, 2006).

15. Wyatt Archeological Museum, Home Page, http://www.wyattmuseum.com (accessed January 13, 2006). *Christian Answers Network Bible Encyclopedia*, s.v. "Ark of the Covenant, Lost or Found?" (by Gary Byers) http://christiananswers.net/q-abr/abr-a002.html (accessed January 18, 2006). The *Christian Answers Network Bible Encyclopedia* has a number of pages that dis-

cuss the Ark of the Covenant; simply use the Web site's search engine to search for "Ark of the Covenant."

16. *Wikipedia Encyclopedia Online*, s.v. "Indian Ocean earthquake," http://en.wikipedia.org/wiki/2004_Indian_Ocean_earthquake (accessed January 18, 2006).

17. WebMagic's Global Earthquake Response Center, "South Asia Earthquake Relief & Recovery," *WebMagic's Global Earthquake Response Center* http://www.earthquake.com.pk/ (accessed January 18, 2006).

18. *Wikipedia Encyclopedia Online*, s.v. "Atlantic Hurricane Season," http://en.wikipedia.org/wiki/2005_Atlantic_hurricane_season (accessed January 2006).

Chapter 5

1. "Introduction to book of Daniel," *The NIV Study Bible* (Grand Rapids, Michigan: Zondervan Bible Publishers, 1985), 1298. The page number to this reference may be subject to change, since the *NIV Study Bible* undergoes successive reprints, but the introduction to the book of Daniel should always be easy to find using the table of contents.

2. George Whitten, "Could the United States be mentioned in Prophecy?" *Christian News Service—Worthy News—Christian Magazine*, http://www.worthynews.com/news-features/united-states-prophecy.html (accessed January 18, 2006); James Lloyd, "The Lion, The Bear & The Leopard," *Christian Media Network Bible Encyclopedia*, http://www.christianmedianetwork.com/lbl2.html (accessed January 18, 2006).

3. Dr. Neil Goldberg, "Rome Project," *The Dalton School*, http://intranet.dalton.org/groups/Rome/ (accessed January 19, 2006).

4. Franco Cavazzi, Illustrated History of the Roman Empire, http://www.roman-empire.net (accessed January 15, 2006).

5. Terri Koontz, B.S. and others, "Chapter 15, A New Political Order in Europe," *World Studies for Christian Schools, Second Edition* (Greenville, South Carolina: Bob Jones University Press, 2000), 424–439.

6. Joe Brankin, The Catholic Missionary Union of England and Wales, http://www.cmu.org.uk/religions/12catholics/cat_pg28.htm (accessed January 19, 2006).

7. Stephen Wagner, "The Fatima Prophecies," *About.com/Paranormal Phenomena*, http://paranormal.about.com/library/weekly/aa070300a.htm (accessed January 19, 2006).

8. Ibid, 2.

9. Central Intelligence Agency (CIA), "Lebanon," *CIA: The World Fact Book*, http://www.cia.gov/cia/publications/factbook/geos/le.html (accessed January 19, 2006).

10. General Douglas MacArthur, *New York Times*, Oct 9, 1955, quoted in Michael S. Heiser, *The Facade* (Superior, Colorado: Superior Books, 2001), 14.

11. President Ronald Reagan, address to the United Nations, September 1987, quoted in Michael S. Heiser, *The Facade* (Superior, Colorado: Superior Books, 2001), 152.

12. Mikhail Gorbachev, Soviet Youth, May 4, 1990, quoted in Michael S. Heiser, *The Facade* (Superior, Colorado: Superior Books, 2001), 162.

13. Fenis Dake, *God's Plan for Man* (Lawrenceville, Georgia: Dake Publishing Inc., 1977), 818–819. If the page number isn't relevant to the text, refer to the index, and look up the phrase "Mystery Babylon."

14. Ibid.

15. Ibid.

16. The Middle East Media Research Institute, Anti-Semitism Documentation Project, http://www.memri.org/antisemitism.html (accessed January 19, 2006).

Chapter 6

1. *The Columbia Encyclopedia, Sixth Edition*, s.v. "Satellite, Artificial," http://www.encyclopedia.com/html/section/satelart_SatelliteOrbits.asp (accessed January 2006).

2. Pixar Animation Studios animators, "Commentaries," Disk 2. Finding Nemo, DVD (Emeryville, California: Walt Disney Pictures/Pixar Animation Studios production, 2001).

978-0-595-37238-6
0-595-37238-4

Made in the USA
Lexington, KY
27 November 2010